AF695708

Chi-squared Goodness-of-fit Tests for Censored Data

Stochastic Models in Survival Analysis and Reliability Set

coordinated by
Catherine Huber-Carol and Mikhail Nikulin

Volume 3

Chi-squared Goodness-of-fit Tests for Censored Data

Mikhail S. Nikulin
Ekaterina V. Chimitova

iSTE

WILEY

First published 2017 in Great Britain and the United States by ISTE Ltd and John Wiley & Sons, Inc.

ISTE Ltd
27-37 St George's Road
London SW19 4EU
UK

www.iste.co.uk

John Wiley & Sons, Inc.
111 River Street
Hoboken, NJ 07030
USA

www.wiley.com

Library of Congress Control Number: 2017934786

British Library Cataloguing-in-Publication Data
A CIP record for this book is available from the British Library
ISBN 978-1-78630-000-3

MIX
Paper from responsible sources
FSC www.fsc.org FSC® C013604

Contents

Introduction

Chi-squared testing is one of the most commonly applied statistical techniques. The chi-squared test and its modifications are adopted for testing goodness-of-fit of various probabilistic models and provide reliable answers for researchers in engineering, quality control, finance, medicine, biology and other fields.

This book is devoted to the problems of construction and application of chi-squared goodness-of-fit tests for complete and censored data. We give examples of chi-squared tests for various distributions widely used in practice, and also consider chi-squared tests for parametric proportional hazards models and accelerated failure time models, which are widely used in reliability and survival analysis. Special attention is paid to the choice of grouping intervals and simulations.

Unfortunately, in many books on statistical data analysis, chi-squared tests are applied incorrectly or inefficiently. Classical chi-squared tests assume that unknown distribution parameters are estimated using grouped data, but in practice this assumption is often forgotten. In this book, we consider the modified chi-squared tests, which do not suffer from such drawback. Moreover, we describe the chi-squared goodness-of-fit tests for censored data with time-dependent covariates. Hence, the aim of this book is to demonstrate the application of the modified chi-squared tests in a wide variety of specific situations.

In the complete data case, a well-known modification of the classical chi-squared tests is the Nikulin-Rao-Robson statistic, which is based on the differences between two estimators of the probabilities to fall into grouping intervals: one estimator is based on the empirical distribution function, the other on the maximum likelihood estimators of unknown parameters of the tested model using initial non-grouped data [NIK 73b], [NIK 73a], [NIK 73c], [DZH 74], [KAP 58], [DZH 82], [DZH 83], [RAO 74], [VOI 93a], [VOI 96], [GRE 96], [DRO 88], [DRO 89], [VAN 98], [KEC 02], [VOI 13].

Li and Doss [LI 93] consider the modification of these results using wider classes of estimators from initial data. Goodness-of-fit tests for linear regression have been studied in [MUK 77], [PIE 79], [LOY 80], [KOU 84], [ZAC 71] and [VOI 07].

Habib and Thomas [HAB 86] and Hollander and Peña [HOL 92] considered natural modifications of the Nikulin-Rao-Robson statistic in the case of censored data without covariates. These tests are also based on the differences between two estimators of the probabilities to fall into grouping intervals. The first estimator is based on the Kaplan-Meier estimator of the cumulative distribution function, the second estimator is based on the maximum likelihood estimators of unknown parameters of the tested model using initial non-grouped censored data. Nikulin and Solev [NIK 99] considered the Pearson-type chi-squared goodness-of-fit test for doubly censored data.

The idea of comparing observed and expected numbers of failures in time intervals is due to Akritas [AKR 88] and was further developed by Hjort [HJO 90]. In the censored data case, Hjort [HJO 90], Royston & Lambert [ROY 11] and Pena [PEN 98a] considered goodness-of-fit for parametric Cox models, Gray and Pierce [GRA 85], Akritas and Torbeyns [AKR 97] and Zhang [ZHA 99] for linear regression models. Pena [PEN 98b] considered smooth goodness-of-fit tests for composite hypothesis in hazard-based models with possible covariates.

This book presents the recent innovations in the area as well as the important results previously only published in Russian. Different approaches to the choice of grouping intervals are considered. The chi-squared tests are compared with other goodness-of-fit tests (such as the Kolmogorov, Cramer-von Mises-Smirnov, Anderson-Darling and Zhang tests) in terms of power, when testing close competing hypotheses. The considerable contribution to the development of this direction belongs to Lemeshko (see, for example, [LEM 98], [LEM 00], [LEM 15b], [LEM 09c] and [LEM 10b]).

The book is intended for researchers interested in statistical testing on the basis of censored and complete data, as well as for university teachers and students. The book is also useful for representatives of industry and finance, in survival analysis and reliability, who deal with data analysis.

1

Chi-squared Goodness-of-fit Tests for Complete Data

Let $\mathbf{X} = (X_1, X_2, ..., X_n)^T$ be a sample of independent identically distributed random variables, where $X_i \in \mathbf{R}^1$, belongs to the distribution defined by the cumulative distribution function:

$$F(t) = \mathbf{P}\{X_i < t\}. \qquad [1.1]$$

We can say also that we have *complete data* $X_1, X_2, ..., X_n$.

In this chapter, we consider the problem of testing a composite goodness-of-fit hypothesis:

$$H_0 : F(t) \in \{F(t;\boldsymbol{\theta}), \boldsymbol{\theta} \in \Theta \subset \mathbf{R}^s\}. \qquad [1.2]$$

If the value of parameter $\boldsymbol{\theta}$ is known from prior experiments, then H_0 is a simple hypothesis.

1.1. Classical Pearson's chi-squared test

Now we consider a *partition* of the real line by the points a_i

$$-\infty = a_0 < a_1 < ... < a_{k-1} < a_k = +\infty$$

into k $(k > s+2)$ intervals $(a_{j-1}, a_j]$ for *grouping the data*. Following Cramer [CRA 46], we suppose that:

1) There exists a positive number c such that:

$$p_j(\boldsymbol{\theta}) = \mathbf{P}\{X_1 \in (a_{j-1}, a_j]\} > c > 0,\ j = 1, ..., k;$$

$$p_1(\boldsymbol{\theta}) + p_2(\boldsymbol{\theta}) + ... + p_k(\boldsymbol{\theta}) \equiv 1,\ \boldsymbol{\theta} \in \Theta;$$

2) The functions $p_j(\boldsymbol{\theta})$ have continuous first- and second-order partial derivatives on the set Θ;

3) The matrix

$$\mathbf{B} = \left\| \frac{1}{\sqrt{p_j(\boldsymbol{\theta})}} \frac{\partial p_j(\boldsymbol{\theta})}{\partial \boldsymbol{\theta}_l} \right\|_{k \times s} \qquad [1.3]$$

has rank s.

Furthermore, let $\boldsymbol{\nu} = (\nu_1, ..., \nu_k)^T$ be the result of grouping the random variables $X_1, ..., X_n$ into the intervals $(a_0, a_1], (a_1, a_2], ...,$ (a_{k-1}, a_k). So, instead of complete data $X_1, X_2, ..., X_n$ we have *grouped data* $\nu_1, ..., \nu_k$, where ν_j, $j = 1, ..., k$ are the numbers of observations, fallen into the intervals. As is well known (see, for example, [CRA 46, VOI 93a, VOI 93b]), in the case when H_0 is true and the true value of the parameter $\boldsymbol{\theta}$ is known, *the standard Pearson statistic*:

$$X_n^2(\boldsymbol{\theta}) = \sum_{j=1}^{k} \frac{[\nu_j - np_j(\boldsymbol{\theta})]^2}{np_j(\boldsymbol{\theta})} \qquad [1.4]$$

has in the limit as $n \to \infty$ *the chi-square distribution with* $(k - 1)$ *degrees of freedom.*

The well-known *chi-square criterion of Pearson*, based on the statistic $X_n^2(\boldsymbol{\theta})$, is constructed on this fact. According to this test, *the simple hypothesis*:

$$H_0 : X_i \sim F(t; \boldsymbol{\theta}) \quad \text{is rejected if} \quad X_n^2 > c_\alpha,$$

where $c_\alpha = \chi^2_{k-1,\alpha}$ is the α*-upper quantile* of the chi-square distribution with $(k - 1)$ degrees of freedom.

If $\boldsymbol{\theta}$ is unknown, in this situation we need to estimate it using the data. Hence, the limit distribution of the statistic $X_n^2(\boldsymbol{\theta}_n^*)$ depends on the asymptotical properties of estimator $\boldsymbol{\theta}_n^*$, which we put in [1.4] instead of the unknown parameter $\boldsymbol{\theta}$.

1.2. Joint distribution of $\mathbf{X}_n(\theta_n^*)$ and $\sqrt{n}(\theta_n^* - \theta)$

We denote:

$$\mathbf{X}_n(\boldsymbol{\theta}) = \left(\frac{\nu_1 - np_1(\boldsymbol{\theta})}{\sqrt{np_1(\boldsymbol{\theta})}}, ..., \frac{\nu_k - np_k(\boldsymbol{\theta})}{\sqrt{np_k(\boldsymbol{\theta})}}\right)^T,$$

then, the statistic of Pearson can be written as the square of the length of the vector $\mathbf{X}_n(\boldsymbol{\theta})$:

$$X_n^2(\boldsymbol{\theta}) = \sum_{j=1}^{k} \frac{[\nu_j - np_j(\boldsymbol{\theta})]^2}{np_j(\boldsymbol{\theta})} = \mathbf{X}_n^T(\boldsymbol{\theta})\mathbf{X}_n(\boldsymbol{\theta}).$$

As is known, [CRA 46], [GRE 96], if $n \to \infty$, then under H_0 the vector $\mathbf{X}_n(\boldsymbol{\theta})$ is asymptotically normally distributed:

$$\mathcal{L}\left(\mathbf{X}_n(\boldsymbol{\theta})\right) \to N(\mathbf{0}_k, \mathbf{D}), \qquad [1.5]$$

where:

$$\mathbf{D} = \mathbf{E}_k - \mathbf{q}\mathbf{q}^T, \quad \mathbf{q} = \left(\sqrt{p_1(\boldsymbol{\theta})}, ..., \sqrt{p_k(\boldsymbol{\theta})}\right)^T, \qquad [1.6]$$

where $\mathbf{E}_k$ is the unit matrix of rank k, the matrix $\mathbf{D}$ is idempotent and *rank* $\mathbf{D} = k - 1$.

Furthermore, we denote by C^* the class of all asymptotically normal $\sqrt{n}$-consistent sequences of estimators $\{\boldsymbol{\theta}_n^*\}$, i.e. if $\{\boldsymbol{\theta}_n^*\} \in C^*$, then

$$\mathcal{L}(\sqrt{n}(\boldsymbol{\theta}_n^* - \boldsymbol{\theta})) \to N(\mathbf{0}_s, \mathbf{W}), \qquad n \to \infty,$$

where $\mathbf{W} = \mathbf{W}(\boldsymbol{\theta})$ is a *non-degenerate matrix* of covariance.

Let $\boldsymbol{\theta}$ be unknown and let $\boldsymbol{\theta}_n^*$ be an estimator of the parameter $\boldsymbol{\theta}$, such that as $n \to \infty$

$$\mathcal{L}(\sqrt{n}(\boldsymbol{\theta}_n^* - \boldsymbol{\theta})) \to N(\mathbf{0}_s, \mathbf{W}), \quad \textit{rank}\ \mathbf{W} = s, \qquad [1.7]$$

i.e. $\{\boldsymbol{\theta}_n^*\} \in C^*$.

Evidently, under these suppositions from [1.5] and [1.6], it follows that under H_0, as $n \to \infty$,

$$\mathcal{L}\begin{pmatrix} \mathbf{X}_n(\boldsymbol{\theta}) \\ \sqrt{n}(\boldsymbol{\theta}_n^* - \boldsymbol{\theta}) \end{pmatrix} \to N(\mathbf{0}_{k+s}, \mathbf{C}), \qquad [1.8]$$

where:

$$\mathbf{C} = \begin{Vmatrix} \mathbf{D} & \vdots & \mathbf{Q} \\ \cdots & & \cdots \\ \mathbf{Q}^T & \vdots & \mathbf{W} \end{Vmatrix},$$

$\mathbf{Q} = \|q_{jl}\|_{k \times s}$ is the limit matrix of covariance of the joint distribution of $\mathbf{X}_n(\boldsymbol{\theta})$ and $\sqrt{n}(\boldsymbol{\theta}_n^* - \boldsymbol{\theta})$,

$$q_{jl} = \lim_{n \to \infty} \mathbf{E}\left(\frac{\nu_j - np_j(\boldsymbol{\theta})}{\sqrt{np_j(\boldsymbol{\theta})}} \sqrt{n}(\theta_{nl}^* - \theta_l) \right), \qquad [1.9]$$

and $\boldsymbol{\theta}_n^* = (\theta_{n1}^*, ..., \theta_{ns}^*)^T$. From [1.6] and [1.9], it follows that

$$\mathbf{Q}^T \mathbf{q} = \mathbf{0}_s. \qquad [1.10]$$

We remark that from the conditions $1) - 3)$ of Cramer and from [1.7], it follows that as $n \to \infty$

$$\mathbf{X}_n(\boldsymbol{\theta}_n^*) = \mathbf{X}_n(\boldsymbol{\theta}) - \sqrt{n}\mathbf{B}(\boldsymbol{\theta}_n^* - \boldsymbol{\theta}) + o(\mathbf{1}_s), \qquad [1.11]$$

where the matrix $\mathbf{B}$ is given by [1.3] and $\mathbf{B}^T \mathbf{q} = \mathbf{0}_s$. For this reason from [1.7], [1.8] and [1.11], it follows immediately that under H_0 as $n \to \infty$

$$\mathcal{L}\begin{pmatrix} \mathbf{X}_n(\boldsymbol{\theta}_n^*) \\ \sqrt{n}(\boldsymbol{\theta}_n^* - \boldsymbol{\theta}) \end{pmatrix} \to N(\mathbf{0}_{k+s}, \mathbf{F}), \qquad [1.12]$$

where:

$$\mathbf{F} = \mathbf{F}(\boldsymbol{\theta}) = \begin{Vmatrix} \mathbf{G} & \vdots & \mathbf{Q} - \mathbf{B}\mathbf{W} \\ \cdots & \cdots & \cdots \\ \mathbf{Q}^T - \mathbf{W}\mathbf{B}^T & \vdots & \mathbf{W} \end{Vmatrix}, \qquad [1.13]$$

and

$$\mathbf{G} = \mathbf{E}_k - \mathbf{q}\mathbf{q}^T - \mathbf{Q}\mathbf{B}^T - \mathbf{B}\mathbf{Q}^T + \mathbf{B}\mathbf{W}\mathbf{B}^T. \qquad [1.14]$$

Formulas [1.12]–[1.14] provide necessary material to seek the properties of the criteria of chi-square type.

1.3. Parameter estimation based on complete data. Lemma of Chernoff and Lehmann

We assume that for each $\boldsymbol{\theta} \in \Theta$, the measure $\mathbf{P}_{\boldsymbol{\theta}}$ is absolutely continuous with respect to certain σ-finite measure μ, given on Borelian σ-algebra $\mathcal{B}$. We denote

$$f(t;\boldsymbol{\theta}) = \frac{d\mathbf{P}_{\boldsymbol{\theta}}}{d\mu}(t), \qquad |\, t \,| < \infty,$$

the density of the probability distribution $\mathbf{P}_{\boldsymbol{\theta}}$ with respect to the measure μ (in the continuous case, we assume that μ is the measure of Lebesgue on $\mathcal{B}$ and in the discrete case, μ is the counting measure on $\{0,1,2,...\}$). We denote

$$L_n(\boldsymbol{\theta}) = \prod_{i=1}^{n} f(X_i;\boldsymbol{\theta})$$

the *likelihood function* of the sample $\mathbf{X} = (X_1, ..., X_n)^T$. Concerning the family $\{f(t;\boldsymbol{\theta})\}$, we assume that for sufficiently large n ($n \to \infty$), the conditions of LeCam of the *local asymptotic normality* and *asymptotic differentiability* of the likelihood function $L_n(\boldsymbol{\theta})$ in the point $\boldsymbol{\theta}$ are satisfied (see, for example [LEC 56, IBR 81]):

1) $$L_n(\boldsymbol{\theta} + \frac{1}{\sqrt{n}}\mathbf{h}) - L_n(\boldsymbol{\theta})$$
$$= \frac{1}{\sqrt{n}}\mathbf{h}^T\boldsymbol{\Lambda}_n(\boldsymbol{\theta}) - \frac{1}{2}\mathbf{h}^T\mathbf{I}^{-1}(\boldsymbol{\theta})\mathbf{h} + o_p(1), \qquad \mathbf{h} \in \mathbf{R}^s,$$

2) $$\mathcal{L}\left(\frac{1}{\sqrt{n}}\boldsymbol{\Lambda}_n(\boldsymbol{\theta})\right) \to N_s(\mathbf{0}_s, \mathbf{I}^{-1}(\boldsymbol{\theta})),$$

3) for any $\sqrt{n}$- consistent sequence of estimators $\{\boldsymbol{\theta}_n^*\}$ of the parameter $\boldsymbol{\theta}$,

$$\frac{1}{\sqrt{n}}\left[\boldsymbol{\Lambda}_n(\boldsymbol{\theta}_n^*)-\boldsymbol{\Lambda}_n(\boldsymbol{\theta})\right]=\sqrt{n}\mathbf{I}(\boldsymbol{\theta})(\boldsymbol{\theta}_n^*-\boldsymbol{\theta})+o(\mathbf{1}_s),$$

where:

$$\boldsymbol{\Lambda}_n(\boldsymbol{\theta})=\text{grad }\ln L_n(\boldsymbol{\theta})$$

is the *vector-informant*, based on the sample $\mathbf{X}=(X_1,...,X_n)^T$, $\mathbf{1}_s=(1,...,1)^T$ is the unit vector in $\mathbf{R}^s$, $\mathbf{0}_s$ is the zero vector in $\mathbf{R}^s$ and $\to$ denotes convergence in distribution. Here

$$\mathbf{I}(\boldsymbol{\theta})=\frac{1}{n}\mathbf{E}_{\boldsymbol{\theta}}\boldsymbol{\Lambda}_n(\boldsymbol{\theta})\boldsymbol{\Lambda}_n^T(\boldsymbol{\theta})$$

is the *information matrix of Fisher*, corresponding to the observation X_1, and it is assumed that $\mathbf{I}(\cdot)$ is continuous on Θ and $\det\mathbf{I}(\boldsymbol{\theta})>0$.

We remark that if the conditions of LeCam 1)-3) are true, then C^* is not empty, since there exists the $\sqrt{n}$*-consistent sequence* of *maximum likelihood estimators* $\{\hat{\boldsymbol{\theta}}_n\}$:

$$L_n(\hat{\boldsymbol{\theta}}_n)=\max_{\boldsymbol{\theta}\in\Theta}L_n(\boldsymbol{\theta}),$$

which satisfy the *likelihood equation*:

$$\boldsymbol{\Lambda}_n=\text{grad }\ln L_n(\boldsymbol{\theta})=\mathbf{0}_s,$$

and for which it holds the relation:

$$\mathcal{L}(\sqrt{n}(\hat{\boldsymbol{\theta}}_n-\boldsymbol{\theta}))\to N(\mathbf{0}_s,\mathbf{I}^{-1}). \qquad [1.15]$$

We denote by $C_{\mathbf{I}^{-1}}$ the class of all *asymptotically normal* $\sqrt{n}$*-consistent sequences of estimators* $\{\hat{\boldsymbol{\theta}}_n\}$, which satisfy [1.15]. Evidently, $C_{\mathbf{I}^{-1}}\subset C^*$. If $\{\boldsymbol{\theta}_n^*\}\in C^*$, then from the *LeCam's conditions of regularity* [LEC 60] and from [1.15], it follows that a sequence of estimators:

$$\left\{\boldsymbol{\theta}_n^*+\frac{1}{\sqrt{n}}\mathbf{I}^{-1}(\boldsymbol{\theta}_n^*)\boldsymbol{\Lambda}_n(\boldsymbol{\theta}_n^*)\right\}$$

belongs to the class $C_{\mathbf{I}^{-1}}$, therefore in future an arbitrary element of the class $C_{\mathbf{I}^{-1}}$ will be denoted by $\{\hat{\boldsymbol{\theta}}_n\}$.

LEMMA 1.1 (Chernoff and Lehmann, [CHE 54]).– *Let a random vector* $\mathbf{Y} = (Y_1, ..., Y_k)^T$ *follow the normal distribution* $N_k(\mathbf{0}_k, \mathbf{A})$ *with the parameters:*

$$\mathbf{EY} = \mathbf{0}_k \text{ and } \mathbf{Var}\mathbf{Y} = \mathbf{EYY}^T = \mathbf{A}.$$

Then, the random variable:

$$\mathbf{Y}^T\mathbf{Y} = \sum_{j=1}^{k} Y_j^2$$

is distributed as:

$$\lambda_1\xi_1^2 + \lambda_2\xi_2^2 + ... + \lambda_k\xi_k^2,$$

where $\xi_1, \xi_2, ..., \xi_k$ *are independent standard normal,* $N(0,1)$*, random variables and* $\lambda_1, \lambda_2, ..., \lambda_k$ *are the eigenvalues of the matrix* $\mathbf{A}$.

COROLLARY 1.1.– Let $\mathbf{A}$ be an *idempotent matrix* of order k, i.e.

$$\mathbf{A}^2 = \mathbf{A}^T\mathbf{A} = \mathbf{A}\mathbf{A}^T = \mathbf{A},$$

then, all its eigenvalues λ_i are 0 or 1 and hence *rank* $\mathbf{A} = tr\mathbf{A} = m \leq k$, where m is the number of λ_i which are equal to 1. In this case:

$$\mathbf{P}\left\{\mathbf{Y}^T\mathbf{Y} < x\right\} = \mathbf{P}\left\{\chi_m^2 < x\right\}.$$

COROLLARY 1.2.– Let $\mathbf{B}$ be a symmetric matrix of order k. Then, the random variable $\mathbf{Y}^T\mathbf{B}\mathbf{Y}$ is distributed as:

$$\sum_{j=1}^{k} \mu_j Y_j^2,$$

where $\mu_1, \mu_2, ..., \mu_k$ are the eigenvalues of the matrix $\mathbf{AB}$.

Let $\hat{\boldsymbol{\theta}}_n \in C_{\mathbf{I}^{-1}}$, i.e.

$$\mathcal{L}(\sqrt{n}(\hat{\boldsymbol{\theta}}_n - \boldsymbol{\theta})) \to N(\mathbf{0}_s, \mathbf{I}^{-1}), \qquad n \to \infty.$$

In this case, we have:

$$\mathcal{L}\begin{pmatrix} \mathbf{X}_n(\hat{\boldsymbol{\theta}}_n) \\ \sqrt{n}(\hat{\boldsymbol{\theta}}_n - \boldsymbol{\theta}) \end{pmatrix} \to N\left(\mathbf{0}_{k+s}, \left\| \begin{matrix} \mathbf{G} & \vdots & \mathbf{0}_{k\times s} \\ \cdots & \cdots & \cdots \\ \mathbf{0}_{s\times k} & \vdots & \mathbf{I}^{-1} \end{matrix} \right\|\right), \quad [1.16]$$

$$\mathbf{G} = \mathbf{E}_k - \mathbf{q}\mathbf{q}^T - \mathbf{B}\mathbf{I}^{-1}\mathbf{B}^T, \; \mathbf{q} = \left(\sqrt{p_1(\boldsymbol{\theta})}, ..., \sqrt{p_k(\boldsymbol{\theta})}\right)^T, \quad [1.17]$$

the matrix $\mathbf{B}$ is given by [1.3], *rank* $\mathbf{G} = k - 1$.

As was shown by Chernoff and Lehmann [CHE 54], s eigenvalues, for example $\lambda_1, ..., \lambda_s$, of $\mathbf{G}$ are the roots of the equation:

$$\mid (1-\lambda)\mathbf{I}(\boldsymbol{\theta}) - \mathbf{J}(\boldsymbol{\theta}) \mid = 0,$$

such that $0 < \lambda_i < 1$ $(i = 1, 2, ..., s)$, one eigenvalue, for example λ_{s+1}, is equal to 0 and the other $k - s - 1$ eigenvalues are equal to 1. Here, $n\mathbf{J}(\boldsymbol{\theta})$ is the information matrix of Fisher of the statistic $\boldsymbol{\nu} = (\nu_1, ..., \nu_k)^T$.

Hence, from the Chernoff-Lehmann lemma it follows that:

$$\lim_{n\to\infty} \mathbf{P}\left\{X_n^2(\hat{\boldsymbol{\theta}}_n) < x \mid H_0\right\} = \mathbf{P}\left\{\chi^2_{k-s-1} + \lambda_1\xi_1^2 + ... + \lambda_s\xi_s^2 < x\right\}, \quad [1.18]$$

where $\chi^2_{k-s-1}, \xi_1^2, ..., \xi_s^2$ are the independent random variables, $\xi_i \sim N(0, 1)$, and in general $\lambda_i = \lambda_i(\boldsymbol{\theta}), 0 < \lambda_i < 1$.

EXAMPLE 1.1.– We wish to test the hypothesis H_0 according to which $X_i \sim N(\mu, \sigma^2)$. In this case, there exist the minimal sufficient statistic $\hat{\boldsymbol{\theta}}_n = (\bar{X}_n, s_n^2)^T$,

$$\bar{X}_n = \frac{1}{n}\sum_{i=1}^n X_i \text{ and } s_n^2 = \frac{1}{n}\sum_{i=1}^n (X_i - \bar{X}_n)^2,$$

which is the maximum likelihood estimator (at the same time, $\hat{\boldsymbol{\theta}}_n$ is the method of moments estimator) for $\boldsymbol{\theta} = (\mu, \sigma^2)^T$, and consequently, using the estimators $\hat{\boldsymbol{\theta}}_n = (\bar{X}_n, s_n^2)^T$ of parameter $\boldsymbol{\theta} = (\mu, \sigma^2)^T$, we obtain the result of Chernoff and Lehmann according to which

$$\lim_{n\to\infty} \mathbf{P}\left\{X_n^2(\hat{\boldsymbol{\theta}}_n) < x \mid H_0\right\} = \mathbf{P}\left\{\chi_{k-3}^2 + \lambda_1\xi_1^2 + \lambda_2\xi_2^2 < x\right\},$$

where λ_1 and λ_2 are given by [1.18].

In Figure 1.1, there is the distribution of the statistic $X_n^2(\hat{\boldsymbol{\theta}}_n)$, which was obtained by Monte Carlo simulations. Samples of size $n = 200$ were generated from the normal distribution with the parameters $\mu = 0$ and $\sigma^2 = 1$. Pearson's statistic $X_n^2(\hat{\boldsymbol{\theta}}_n)$ was calculated by the grouped samples with the following boundary points:

$$a_0 = -\infty, \quad a_1 = -1, \quad a_2 = 0, \quad a_3 = 1, \quad a_4 = \infty, \quad k = 4.$$

The maximum likelihood estimates $\hat{\boldsymbol{\theta}}_n$ were calculated by the complete samples. The χ^2 distribution with $(k - s - 1) = 1$ degree of freedom is also given for comparison in Figure 1.1.

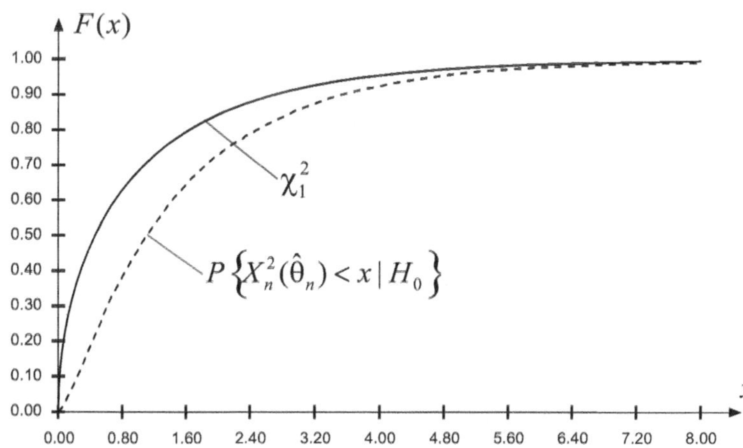

Figure 1.1. *The distribution of the statistic $X_n^2(\hat{\boldsymbol{\theta}}_n)$*

As can be seen from this figure, if the maximum likelihood estimate of the parameter $\boldsymbol{\theta}$ is obtained from the initial non-grouped data, then the distribution of Pearson's statistic under H_0 significantly differs from the chi-square distribution with $(k - s - 1)$ degrees of freedom.

EXAMPLE 1.2.– We wish to test the hypothesis H_0 according to which

$$X_i \sim \frac{\theta^t}{t!} exp(-\theta),\ t = 0, 1, 2, ...;\ \theta > 0,$$

i.e. under H_0 the random variable X_i follows the law of Poisson with the parameter θ. In this case, the statistic $\hat{\theta}_n = \bar{X}_n$ is the maximum likelihood estimator (at the same time, $\hat{\theta}_n$ is the method of moments estimator) of parameter θ, and hence from the lemma of Chernoff and Lehmann:

$$\lim_{n\to\infty} \mathbf{P}\left\{X_n^2(\hat{\theta}_n) < x \mid H_0\right\} = \mathbf{P}\left\{\chi^2_{k-2} + \lambda_1\xi_1^2 < x\right\},$$

where λ_1 is given by [1.18].

1.4. Parameter estimation based on grouped data. Theorem of Fisher

We denote by $C_{\mathbf{J}^{-1}}$ the class of all sequences $\{\tilde{\boldsymbol{\theta}}_n\}$ of estimators $\tilde{\boldsymbol{\theta}}_n$, which satisfy the condition:

$$\mathcal{L}(\sqrt{n}(\tilde{\boldsymbol{\theta}}_n - \boldsymbol{\theta})) \to N(\mathbf{0}_s, \mathbf{J}^{-1}), \qquad [1.19]$$

where:

$$n\mathbf{J} = n\mathbf{J}(\boldsymbol{\theta}) = n\mathbf{B}^T\mathbf{B}$$

is the *information matrix of Fisher* of the statistic $\boldsymbol{\nu} = (\nu_1, ..., \nu_k)^T$, i.e. in this case, $\mathbf{W} = \mathbf{J}^{-1}$. Evidently, $C_{\mathbf{J}^{-1}} \subset C^*$. We remark that from [1.11], it follows that:

$$\sqrt{n}(\tilde{\boldsymbol{\theta}}_n - \boldsymbol{\theta}) = \mathbf{J}^{-1}\mathbf{B}^T\mathbf{X}_n(\boldsymbol{\theta}) + o_p(\mathbf{1}_s), \quad n \to \infty,$$

and the limit matrix of covariance $\mathbf{Q}$ in virtue [1.10] is:

$$\mathbf{Q} = (\mathbf{E}_k - \mathbf{q}\mathbf{q}^T)\mathbf{B}\mathbf{J}^{-1} = \mathbf{B}\mathbf{J}^{-1}, \qquad [1.20]$$

because $\mathbf{q}^T\mathbf{B} = \mathbf{0}_s^T$, and since $\mathbf{Q}^T\mathbf{Q} = \mathbf{J}^{-1}$ we find that:

$$\mathbf{Q} - \mathbf{BW} = \mathbf{BJ}^{-1} - \mathbf{BJ}^{-1} = \mathbf{0}_{k\times s},$$

from which we obtain that under H_0 for any sequence of estimators $\{\tilde{\boldsymbol{\theta}}_n\}$, which satisfies [1.19], we have:

$$\mathcal{L}\begin{pmatrix} \mathbf{X}_n(\tilde{\boldsymbol{\theta}}_n) \\ \sqrt{n}(\tilde{\boldsymbol{\theta}}_n - \boldsymbol{\theta}) \end{pmatrix} \to N(\mathbf{0}_{k+s}, \tilde{\mathbf{C}}), \qquad n \to \infty \qquad [1.21]$$

where:

$$\tilde{\mathbf{C}} = \begin{Vmatrix} \mathbf{G} & \vdots & \mathbf{0}_{k\times s} \\ \cdots & \cdots & \cdots \\ \mathbf{0}_{k\times s} & \vdots & \mathbf{J}^{-1} \end{Vmatrix}, \qquad \mathbf{G} = \mathbf{E}_k - \mathbf{q}\mathbf{q}^T - \mathbf{BJ}^{-1}\mathbf{B}^T, \ [1.22]$$

rank $\mathbf{G} = k - s - 1$ and the matrix $\mathbf{G}$ is idempotent. We remark that from [1.21], it follows the asymptotical independence of $\mathbf{X}_n(\tilde{\boldsymbol{\theta}})$ and $\sqrt{n}(\tilde{\boldsymbol{\theta}}_n - \boldsymbol{\theta})$ (see, for example, [CRA 46]). Finally, from [1.21], [1.22] and Corollary 2 of the Chernoff-Lehmann lemma, we obtain *the theorem of Fisher*:

$$\lim_{n\to\infty} \mathbf{P}\left\{X_n^2(\tilde{\boldsymbol{\theta}}_n) < x \mid H_0\right\} = \mathbf{P}\left\{\chi^2_{k-s-1} < x\right\}.$$

So, we see that if grouped data are used for estimation of unknown parameter $\boldsymbol{\theta}$, then the limit distribution of the obtained statistic $X_n^2(\tilde{\boldsymbol{\theta}}_n)$ is chi-square with $(k - s - 1)$ degrees of freedom.

Let us consider several examples of such estimators:

1) As is known (see, for example, [CRA 46]), the *estimator of Fisher* $\tilde{\boldsymbol{\theta}}_n$, obtained by the *minimum chi-squared method*

$$X_n^2(\tilde{\boldsymbol{\theta}}) = \min_{\boldsymbol{\theta}\in\Theta} X_n^2(\boldsymbol{\theta}),$$

satisfies the relation [1.20].

2) The minimum chi-squared method can be modified by replacing the denominator of [1.4] with ν_j, which simplifies the calculations. This method is referred to as *modified minimum chi-squared method.*

3) As shown by Cramer [CRA 46], the *multinomial maximum likelihood estimator* $\tilde{\boldsymbol{\theta}}_n$ - the maximum point of the likelihood function $l_n(\boldsymbol{\theta})$ of the statistic $\boldsymbol{\nu} = (\nu_1, ..., \nu_k)^T$:

$$l_n(\tilde{\boldsymbol{\theta}}_n) = \max_{\boldsymbol{\theta} \in \Theta} l_n(\boldsymbol{\theta}), \; l_n(\boldsymbol{\theta}) = p_1^{\nu_1}(\boldsymbol{\theta}) p_2^{\nu_2}(\boldsymbol{\theta}) ... p_k^{\nu_k}(\boldsymbol{\theta}),$$

also satisfies the relation [1.20].

We note here that if $X_1, ..., X_n$ follow a continuous distribution $f(t; \boldsymbol{\theta})$, $\boldsymbol{\theta} \in \Theta$, then the statistic $\boldsymbol{\nu} = (\nu_1, ..., \nu_k)^T$ is *not sufficient* like the vector of observations $\mathbf{X} = (X_1, ..., X_n)^T$ itself, and hence in this case, the matrix

$$n\mathbf{I}(\boldsymbol{\theta}) - n\mathbf{J}(\boldsymbol{\theta}) \qquad [1.23]$$

is positively definite, where $n\mathbf{I}(\boldsymbol{\theta})$ is the *information matrix of the sample* $\mathbf{X} = (X_1, ..., X_n)^T$.

1.5. Nikulin-Rao-Robson chi-squared test

When testing goodness-of-fit of a continuous distribution on the basis of complete data, the classical Pearson chi-squared test has strong drawbacks:

1) Asymptotically optimal estimators (such as maximum likelihood estimators for regular models) from the initial complete data cannot be used in Pearson's statistic, because the limit distribution of Pearson's statistic depends on unknown parameters.

2) The estimators $\tilde{\boldsymbol{\theta}}$ from grouped data are not optimal, because they do not use all data, hence we lose the power of goodness-of-fit tests, based on such estimators. Moreover, their computing requires solution of complicated systems of equations.

Here, we consider the modification of Pearson's statistic, which does not have the above-mentioned drawbacks. So, let $\mathbf{X} = (X_1, ..., X_n)^T$ be the sample. We want to test the hypothesis H_0 according to which $X_i \sim f(t; \boldsymbol{\theta})$, and let $\hat{\boldsymbol{\theta}}_n$ be the maximum likelihood estimator for $\boldsymbol{\theta}$, for which the relation [1.15] holds. From [1.5]–[1.16], it follows that it is possible to take the statistic:

$$Y_n^2 = Y_n^2(\hat{\boldsymbol{\theta}}_n) = \mathbf{X}_n^T(\hat{\boldsymbol{\theta}}_n)\mathbf{G}^-(\hat{\boldsymbol{\theta}}_n)\mathbf{X}_n(\hat{\boldsymbol{\theta}}_n), \qquad [1.24]$$

where the matrix $\mathbf{G}$ is given by [1.17]. We note that the quadratic form Y_n^2 is invariant under choice of $\mathbf{G}^-$ in virtue of the specific condition

$$p_1(\boldsymbol{\theta}) + p_2(\boldsymbol{\theta}) + ... + p_k(\boldsymbol{\theta}) \equiv 1$$

of the singularity of the matrix $\mathbf{G}$. As follows from the lemma of Chernoff and Lehmann:

$$\lim \mathbf{P}\left\{Y_n(\hat{\boldsymbol{\theta}}_n) < x \mid H_0\right\} = \mathbf{P}\left\{\chi^2_{k-1} < x\right\}. \qquad [1.25]$$

Some examples of the use of statistic Y_n^2 can be found in the papers of Dudley [DUD 79], Nikulin & Voinov [NIK 89], [NIK 90b], Nikulin [NIK 73b], [NIK 73a], [NIK 73c], [NIK 79b], [NIK 79a], [NIK 90a], [NIK 74], [NIK 91], Greenwood & Nikulin [GRE 88], [GRE 96], Bolshev & Nikulin [BOL 75], Bolshev & Mirvaliev [BOL 78], Dzhaparidze & Nikulin [DZH 74], [DZH 82], Beinicke & Dzhaparidze [BEI 82], Mirvaliev & Nikulin [MIR 89], Nikulin & Yusas [NIK 83], Chibisov [CHI 71], Lemeshko [LEM 01], Voinov [VOI 06], Bagdonavicius, Kruopis and Nikulin [BAG 10a], where in particular the statistic Y_n^2 is constructed for testing normality, for the problem of homogeneity of two or more samples from the distributions with shift and scale parameters, in the problem when only a part of the sample is used to estimate the unknown parameter, and also some applications including the problem of the construction of a chi-square test with a random cell boundary.

The statistic $Y_n^2(\hat{\boldsymbol{\theta}}_n)$ has a particularly convenient form when we construct a chi-squared test with random cell boundaries for continuous distributions. Let us fix the vector of probability $\mathbf{p} = (p_1, ..., p_k)^T$ such that $p_j > 0$ $(j = 1, 2, ..., k)$ and $\mathbf{p}^T\mathbf{1}_k = 1$, and let $a_1(\hat{\boldsymbol{\theta}}_n), ..., a_{k-1}(\hat{\boldsymbol{\theta}}_n)$ be the boundaries of intervals $(a_{j-1}(\hat{\boldsymbol{\theta}}_n), a_j(\hat{\boldsymbol{\theta}}_n)]$, where $a_j(\hat{\boldsymbol{\theta}}_n)$ are given by:

$$p_j = \mathbf{P}\left\{X_1 \in (a_{j-1}(\hat{\boldsymbol{\theta}}_n), a_j(\hat{\boldsymbol{\theta}}_n)]\right\} = \int_{a_{j-1}(\hat{\boldsymbol{\theta}}_n)}^{a_j(\hat{\boldsymbol{\theta}}_n)} f(x; \hat{\boldsymbol{\theta}}_n)dx,$$

where $a_0(\hat{\boldsymbol{\theta}}_n) = -\infty$, $a_k(\hat{\boldsymbol{\theta}}_n) = +\infty$. We choose the probabilities p_j, independent from $\boldsymbol{\theta}$, and hence we obtain the intervals with the cell boundary. We suppose that $a_j(\boldsymbol{\theta})$ are differentiable functions. The elements of the matrix

$$\mathbf{C} = \left\| \frac{\partial p_j(\boldsymbol{\theta})}{\partial \boldsymbol{\theta}_l} \right\|_{k \times s}$$

are defined as follows:

$$c_{jl} = c_{jl}(\boldsymbol{\theta}) = f(a_{j-1}(\boldsymbol{\theta}); \boldsymbol{\theta}) \frac{\partial a_{j-1}(\boldsymbol{\theta})}{\partial \boldsymbol{\theta}} - f(a_j(\theta); \boldsymbol{\theta}) \frac{\partial a_j(\boldsymbol{\theta})}{\partial \boldsymbol{\theta}},$$

and let $\boldsymbol{\nu} = (\nu_1, ..., \nu_k)^T$ be the result of grouping of $X_1, ..., X_n$ upon the intervals:

$$(a_0(\hat{\boldsymbol{\theta}}_n), a_1(\hat{\boldsymbol{\theta}}_n)], ..., (a_{k-1}(\hat{\boldsymbol{\theta}}_n), a_k(\hat{\boldsymbol{\theta}}_n)).$$

Then, the test statistic [1.24] can be written in the form:

$$Y_n^2(\hat{\boldsymbol{\theta}}_n) = \sum_{j=1}^{k} \frac{(\nu_j - np_j)^2}{np_j} + \frac{1}{n} \mathbf{V}^T \boldsymbol{\Lambda}^{-1}(\boldsymbol{\theta}) \mathbf{V}, \qquad [1.26]$$

where:

$$\mathbf{V} = (v_1, ..., v_s)^T, \quad v_l = \frac{c_{1l}\nu_1}{p_1} + ... + \frac{c_{kl}\nu_k}{p_k},$$

$$\boldsymbol{\Lambda}(\boldsymbol{\theta}) = \mathbf{I}(\boldsymbol{\theta}) - \mathbf{J}(\boldsymbol{\theta}), \quad \mathbf{J}(\boldsymbol{\theta}) = \left\| \frac{c_{jl}c_{jl'}}{p_j} \right\|_{s \times s}.$$

As shown by Nikulin ([NIK 73b]),

$$\lim_{n \to \infty} \mathbf{P} \left\{ Y_n^2(\hat{\theta}_n) < x \mid H_0 \right\} = \mathbf{P} \left\{ \chi_{k-1}^2 < x \right\}.$$

EXAMPLE 1.3 (Testing normality).– Let $X_1, X_2, ..., X_n$ be independent identically distributed random variables. We wish to test the hypothesis H_0, according to which

$$X_i \sim \Phi\left(\frac{t - \mu}{\sigma}\right), \quad |\mu| < \infty, \ \sigma > 0,$$

where $\Phi(t)$ is the distribution function of the standard normal law. The Fisher information matrix is:

$$I(\boldsymbol{\theta}) = \frac{1}{\sigma^2} \left\| \begin{matrix} 1 & 0 \\ 0 & 2 \end{matrix} \right\|, \quad \boldsymbol{\theta} = (\mu, \sigma)^T.$$

The statistics

$$\bar{X}_n = \frac{1}{n} \sum_{i=1}^{n} X_i \quad \text{and} \quad s_n^2 = \frac{1}{n} \sum_{i=1}^{n} (X_i - \bar{X}_n)^2,$$

are the maximum likelihood estimators for the parameters μ and σ^2. Then, set:

$$Y_i = \frac{1}{s_n} \left(X_i - \bar{X}_n \right), \quad i = 1, ..., n.$$

Let $\mathbf{p} = (p_1, ..., p_k)^T$ be the vector of positive probabilities such that:

$$p_1 + ... + p_k = 1, \quad p_j > 0, \quad j = 1, ..., k,$$

and let:

$$x_j = \Phi^{-1}(p_1 + ... + p_j), \quad j = 1, ..., k-1.$$

Setting $x_0 = -\infty$, $x_k = +\infty$, consider $\boldsymbol{\nu} = (\nu_1, ..., \nu_k)^T$, the result of grouping $Y_1, Y_2, ..., Y_n$ using the intervals $(x_0, x_1], (x_1, x_2], ...,$ (x_{k-1}, x_k). It is evident that the same vector $\boldsymbol{\nu}$ is obtained by grouping $X_1, X_2, ..., X_n$ using the intervals:

$$(a_0, a_1], (a_1, a_2], ..., (a_{k-1}, a_k), \quad a_j = x_j s_n + \bar{X}_n, \quad j = 0, 1, ..., k.$$

Hence, the elements of the matrix

$$\mathbf{C} = \left\| \frac{\partial p_j(\boldsymbol{\theta})}{\partial \boldsymbol{\theta}_l} \right\|_{k \times 2}$$

are calculated as follows:

$$c_{j1} = \varphi(x_j) - \varphi(x_{j-1}),$$

$$c_{j2} = -x_j\varphi(x_j) + x_{j-1}\varphi(x_{j-1}),$$

where $\varphi(t)$ is the density function of the standard normal law.

So, the Nikulin-Rao-Robson statistic can be calculated by the following formula:

$$Y_n^2 = X_n^2 + \frac{\lambda_1\beta^2(\boldsymbol{\nu}) - 2\lambda_3\alpha(\boldsymbol{\nu})\beta(\boldsymbol{\nu}) + \lambda_2\alpha^2(\boldsymbol{\nu})}{n\left(\lambda_1\lambda_2 - \lambda_3^2\right)},$$

where:

$$X_n^2 = \sum_{j=1}^{k} \frac{(\nu_j - np_j)^2}{np_j},$$

$$\alpha(\boldsymbol{\nu}) = \sum_{j=1}^{k} \frac{c_{j1}\nu_j}{p_j}, \quad \beta(\boldsymbol{\nu}) = \sum_{j=1}^{k} \frac{c_{j2}\nu_j}{p_j},$$

$$\lambda_1 = 1 - \sum_{j=1}^{k} \frac{c_{j1}^2}{p_j}, \quad \lambda_2 = 2 - \sum_{j=1}^{k} \frac{c_{j2}^2}{p_j}, \quad \lambda_3 = -\sum_{j=1}^{k} \frac{c_{j1}c_{j2}}{p_j}.$$

We know that under H_0 as $n \to \infty$, the statistic Y_n^2 has the chi-square distribution with $(k-1)$ degrees of freedom.

EXAMPLE 1.4 (Exponential distribution).– Let $X_1, X_2, ..., X_n$ be independent identically distributed random variables. We wish to test the hypothesis H_0, according to which

$$X_i \sim F(t;\theta) = 1 - e^{-t/\theta}, \quad t \geq 0, \;\; \theta > 0,$$

meaning that random variables have exponential distribution.

The Fisher information $I(\theta) = 1/\theta$. The maximum likelihood estimator of unknown parameter $\hat{\theta}_n = \bar{X}_n$. For the given vector of probabilities, we have:

$$x_j = -\ln(1 - p_j), \quad a_j = \hat{\theta}_n x_j, \quad c_{j1} = \frac{x_j e^{-x_j} - x_{j-1}e^{-x_{j-1}}}{\hat{\theta}_n} = \frac{b_j}{\hat{\theta}_n}.$$

So, the test statistic is written in the form:

$$Y_n^2 = X_n^2 + \frac{v^2}{n\lambda},$$

where:

$$X_n^2 = \sum_{j=1}^{k} \frac{(\nu_j - np_j)^2}{np_j}, \quad v = \sum_{j=1}^{k} \frac{b_j \nu_j}{p_j}, \quad \lambda = 1 - \sum_{j=1}^{k} \frac{b_j^2}{p_j}.$$

EXAMPLE 1.5 (Cauchy distribution).– We want to test the hypothesis H_0, according to which the independent random variables $X_1, X_2, ..., X_n$ follow the Cauchy distribution with density:

$$f(t;\theta) = \frac{1}{\pi[1+(t-\theta)^2]}, \mid t \mid < \infty, \; \theta \in \mathbf{R}^1.$$

In this case, θ is the scalar parameter and the information of Fisher $I(\theta)$ corresponding to one observation X_1 is:

$$I(\theta) = \mathbf{E}\left\{\frac{\partial \ln f(X_1,\theta)}{\partial \theta}\right\}^2 = \frac{4}{\pi}\int_{-\infty}^{\infty} \frac{(t-\theta)^2}{[1+(t-\theta)^2]^3} dt = \frac{1}{2}.$$

Let us fix the vector $\mathbf{p} = (p_1, ..., p_k)^T$ such that $p_1 = p_2 = ... = p_k = \frac{1}{k}$ and we define the points $x_1, x_2, ..., x_{k-1}$ from the conditions:

$$-\infty = x_0 < x_1 < ... < x_{k-1} < x_k = +\infty,$$

$$\frac{j}{k} = \frac{1}{\pi}\int_{-\infty}^{x_j} \frac{dx}{1+x^2}, \qquad j = 1, ..., k-1.$$

As is known, the maximum likelihood estimator $\hat{\theta}_n$ of the parameter θ in this problem is the root of the equation:

$$\theta = \frac{\sum_{i=1}^{n} \frac{X_i}{1+(X_i-\theta)^2}}{\sum_{i=1}^{n} \frac{1}{1+(X_i-\theta)^2}}.$$

If $\boldsymbol{\nu} = (\nu_1, ..., \nu_k)^T$ is the vector of frequencies obtained by grouping the random variables $X_1, ..., X_n$ upon the intervals:

$$(-\infty, x_1 + \hat{\theta}_n], (x_1 + \hat{\theta}_n, x_2 + \hat{\theta}_n], ..., (x_{k-1} + \hat{\theta}_n, +\infty),$$

i.e. $a_j(\hat{\theta}_n) = x_j + \hat{\theta}_n$. Then, in this case under H_0, the statistic

$$Y_n^2 = \frac{k}{n}\sum_{j=1}^{k}\nu_j^2 - n + \frac{2k^2}{n\mu}\left(\sum_{j=1}^{k}\nu_j\delta_j\right)^2$$

has in the limit as $n \to \infty$ the chi-square distribution with $(k-1)$ degrees of freedom, where:

$$\delta_j = \frac{1}{\pi}\left[\frac{1}{1+x_{j-1}^2} - \frac{1}{1-x_j^2}\right], \; j = 1, ..., k,$$

$$\mu = 1 - 2k\sum_{j=1}^{k}\delta_j^2.$$

In particular, if $k = 4$, then $x_1 = -1$, $x_2 = 0$, $x_3 = 1$,

$$\delta_1 = \delta_2 = \frac{1}{2\pi}, \; \delta_3 = \delta_4 = -\frac{1}{2\pi}, \; \mu = 1 - \frac{8}{\pi^2},$$

and hence:

$$Y_n^2 = \left(\frac{4}{n}\sum_{j=1}^{4}\nu_j^2 - n\right) + \frac{8}{n(\pi^2-8)}(\nu_1 + \nu_2 - \nu_3 - \nu_4)^2,$$

and

$$\lim_{n\to\infty}\mathbf{P}\left\{Y_n^2 < x \mid H_0\right\} = \mathbf{P}\left\{\chi_3^2 < x\right\}.$$

1.6. Other modifications

Comparing the estimators $\tilde{\boldsymbol{\theta}}_n$ and $\hat{\boldsymbol{\theta}}_n$ from the classes $C_{\mathbf{J}-1}$ and $C_{\mathbf{I}-1}$ correspondingly, and using formulas [1.10], [1.11], [1.16] and [1.21], we can write that as $n \to \infty$

$$\mathcal{L}\begin{pmatrix}\sqrt{n}(\hat{\boldsymbol{\theta}}_n - \boldsymbol{\theta}) \\ \sqrt{n}(\tilde{\boldsymbol{\theta}}_n - \boldsymbol{\theta})\end{pmatrix} \to N(\mathbf{0}_{2s}, \mathbf{G}), \qquad [1.27]$$

where

$$\mathbf{G} = \left\| \begin{array}{ccc} \mathbf{I}^{-1}(\boldsymbol{\theta}) & \vdots & \mathbf{I}^{-1}(\boldsymbol{\theta}) \\ \cdots & \cdots & \cdots \\ \mathbf{I}^{-1}(\boldsymbol{\theta}) & \vdots & \mathbf{J}^{-1}(\boldsymbol{\theta}) \end{array} \right\|, \qquad [1.28]$$

because

$$\begin{aligned} \sqrt{n}(\tilde{\boldsymbol{\theta}}_n - \boldsymbol{\theta}) &= \mathbf{J}^{-1}\mathbf{B}^T\mathbf{X}_n(\boldsymbol{\theta}) + o(\mathbf{1}_s), \\ \sqrt{n}(\hat{\boldsymbol{\theta}}_n - \boldsymbol{\theta}) &= \tfrac{1}{\sqrt{n}}\mathbf{I}^{-1}\boldsymbol{\Lambda}_n(\mathbf{X}) + o(\mathbf{1}_s), \end{aligned} \qquad [1.29]$$

and hence:

$$n\mathbf{E}(\tilde{\boldsymbol{\theta}}_n-\boldsymbol{\theta})(\hat{\boldsymbol{\theta}}_n-\boldsymbol{\theta})^T = \frac{1}{\sqrt{n}}\mathbf{J}^{-1}\mathbf{B}^T\left[\mathbf{E}\mathbf{X}_n(\boldsymbol{\theta})\boldsymbol{\Lambda}_n^T(\mathbf{X})\right]\mathbf{I}^{-1}+o_p(\mathbf{1}_{s\times s}) =$$

$$= \mathbf{J}^{-1}\mathbf{B}^T\mathbf{B}\mathbf{I}^{-1} + o_p(\mathbf{1}_{s\times s} = \mathbf{I}^{-1}(\boldsymbol{\theta}) + o_p(\mathbf{1}_{s\times s}),$$

since $\mathbf{E}\mathbf{X}_n(\boldsymbol{\theta})\boldsymbol{\Lambda}_n^T(\mathbf{X}) = \sqrt{n}\mathbf{B}$.

Finally, from [1.27] and [1.28] it follows that as $n \to \infty$

$$n\mathbf{E}(\hat{\boldsymbol{\theta}}_n - \tilde{\boldsymbol{\theta}}_n)(\hat{\boldsymbol{\theta}}_n - \tilde{\boldsymbol{\theta}}_n)^T = \mathbf{J}^{-1} - \mathbf{I}^{-1} + o_p(\mathbf{1}_{s\times s}),$$

and hence, as $n \to \infty$

$$\mathcal{L}(\sqrt{n}(\hat{\boldsymbol{\theta}}_n - \tilde{\boldsymbol{\theta}}_n)) \to N(\mathbf{0}_s, \mathbf{J}^{-1} - \mathbf{I}^{-1}) \qquad [1.30]$$

(for the continuous distributions the matrix $\mathbf{J}^{-1} - \mathbf{I}^{-1}$ is always non-degenerated, as it follows from [1.23]. From [1.30], we obtain that the statistic

$$M_n^2 = n(\hat{\boldsymbol{\theta}}_n - \tilde{\boldsymbol{\theta}}_n)^T[\mathbf{J}^{-1}(\hat{\boldsymbol{\theta}}_n) - \mathbf{I}^{-1}(\tilde{\boldsymbol{\theta}}_n)]^{-1}(\hat{\boldsymbol{\theta}}_n - \tilde{\boldsymbol{\theta}}_n) \qquad [1.31]$$

has in the limit as $n \to \infty$ the chi-square distribution with s degrees of freedom. We remark that from [1.29] and [1.31], it follows that:

$$\begin{aligned} M_n^2 = \mathbf{X}_n^T(\hat{\boldsymbol{\theta}}_n)\mathbf{B}(\boldsymbol{\theta}_n^*)&\left[(\mathbf{I}(\boldsymbol{\theta}_n^*) - \mathbf{J}(\boldsymbol{\theta}_n^*))^{-1}\right. \\ &\left.+\mathbf{J}^{-1}(\boldsymbol{\theta}_n^*)\right]\mathbf{B}^T(\boldsymbol{\theta}_n^*)\mathbf{X}_n(\hat{\boldsymbol{\theta}}_n) + o_p(1), \end{aligned} \qquad [1.32]$$

where $\{\boldsymbol{\theta}_n^*\} \in C^*$.

In the paper of Dzhaparidze and Nikulin [DZH 74], for testing of the hypothesis H_0 the following statistic was proposed

$$W_n^2 = \mathbf{X}_n^T(\boldsymbol{\theta}_n^*)\left[\mathbf{E}_k - \mathbf{B}(\boldsymbol{\theta}_n^*)\mathbf{J}^{-1}(\boldsymbol{\theta}_n^*)\mathbf{B}^T(\boldsymbol{\theta}_n^*)\right]\mathbf{X}_n(\boldsymbol{\theta}_n^*), \quad [1.33]$$

making it possible to use an arbitrary $\sqrt{n}$- consistent sequence of estimators $\{\boldsymbol{\theta}_n^*\} \in C^*$, since:

$$\lim_{n\to\infty} \mathbf{P}\left\{W_n^2(\boldsymbol{\theta}_n^*) > x \mid H_0\right\} = \mathbf{P}\left\{\chi_{k-s-1}^2 > x\right\}.$$

From [1.18], [1.32] and [1.33], it follows that the statistic Y_n^2 is connected with W_n^2 by the next formula:

$$Y_n^2(\hat{\boldsymbol{\theta}}_n) = \mathbf{X}_n^T(\hat{\boldsymbol{\theta}}_n)\left[\mathbf{E}_k - \mathbf{B}(\hat{\boldsymbol{\theta}}_n)\mathbf{I}^{-1}(\hat{\boldsymbol{\theta}}_n)\mathbf{B}^T(\hat{\boldsymbol{\theta}}_n)\right]^{-1}\mathbf{X}_n(\hat{\boldsymbol{\theta}}_n)$$

$$= W_n^2(\hat{\boldsymbol{\theta}}_n) + M_n^2, \quad [1.34]$$

because

$$(\mathbf{E}_k - \mathbf{B}\mathbf{I}^{-1}\mathbf{B}^T)^{-1} = \mathbf{E}_k + \mathbf{B}(\mathbf{I} - \mathbf{J})^{-1}\mathbf{B}^T.$$

REMARK 1.1.– Generally, the method of moments does not give asymptotically efficient estimators from the class $C_{\mathbf{I}^{-1}}$, but gives only $\sqrt{n}$-consistent estimators from the class C^*. Therefore, we recommend using the statistic $W_n^2(\boldsymbol{\theta}_n^*)$ to test H_0 with the help of the chi-square type criterion when we use any one $\sqrt{n}$-consistent estimator, for example the method of moments estimator.

REMARK 1.2.– As we can see from [1.33] and [1.34], the statistic $Y_n^2(\hat{\boldsymbol{\theta}}_n)$ takes into account the difference in information about $\boldsymbol{\theta}$ between two estimators from classes $C_{\mathbf{I}^{-1}}$ and $C^* \setminus C_{\mathbf{I}^{-1}}$.

1.7. The choice of grouping intervals

When using chi-square goodness-of-fit tests, the problem of choosing boundary points and the number of grouping intervals is always urgent, as the power of these tests considerably depends on the grouping method used. In the case of complete samples (without censored observations), this problem was investigated in [VOI 09], [DEN 89], [LEM 98], [LEM 00], [LEM 01], [DEN 04]. In particular, in [LEM 00], the investigation of the power of the Pearson and Nikulin-Rao-Robson tests for complete samples has been carried out for various numbers of intervals and grouping methods. The chi-squared tests for verification of normality were investigated in [LEM 15a], and in [LEM 15b] they were compared with other goodness-of-fit tests and with the special normality tests.

In [DEN 79], it was shown for the first time that asymptotically optimal grouping, for which the loss of the Fisher information from grouping is minimized, enables the power of the Pearson test to be maximized against close competing hypotheses. For example, it is possible to maximize the determinant of the Fisher information matrix for grouped data:

$$\mathbf{J}(\boldsymbol{\theta}) = \sum_{j=1}^{k} \frac{\frac{\partial}{\partial \boldsymbol{\theta}} p_j(\boldsymbol{\theta}) \left(\frac{\partial}{\partial \boldsymbol{\theta}} p_j(\boldsymbol{\theta}) \right)^T}{p_j(\boldsymbol{\theta})},$$

i.e. to solve the problem of D-optimal grouping:

$$\max_{a_1 < \ldots < a_{k-1}} \det\left(\mathbf{J}(\boldsymbol{\theta})\right). \qquad [1.35]$$

In the case of A-optimality criterion, the trace of the information matrix $\mathbf{J}(\boldsymbol{\theta})$ is maximized by the boundary points:

$$\max_{a_1 < \ldots < a_{k-1}} \operatorname{Tr}\left(\mathbf{J}(\boldsymbol{\theta})\right), \qquad [1.36]$$

and E-optimality criterion maximizes the minimum eigenvalue of the information matrix:

$$\max_{a_1 < \ldots < a_{k-1}} \min_{l=\overline{1,q}} \lambda_l\left(\mathbf{J}(\boldsymbol{\theta})\right). \qquad [1.37]$$

The problem of asymptotically optimal grouping (AOG) by the D-optimality criteria was solved by Lemeshko in [LEM 98], and the tables of D-optimal grouping are given in Appendix A for a certain number of distributions. In Table 1.1, we give the density functions and the references to the corresponding tables of AOG.

Distribution	Density function	a_j	Table of AOG x_j	Table of AOG p_j
Exponential	$\frac{1}{\theta}e^{-t/\theta}, \ t \geq 0$	$a_j = \theta x_j$	A.1	A.2
Rayleigh	$\frac{t}{\theta^2}e^{-t^2/2\theta^2}, \ t \geq 0$	$a_j = \theta x_j$	A.3	A.2
Maxwell	$\frac{2t^2}{\theta^3\sqrt{2\pi}}e^{-t^2/2\theta^2}, \ t \geq 0$	$a_j = \theta x_j$	A.4	A.5
Half-normal	$\frac{2}{\theta\sqrt{2\pi}}e^{-t^2/2\theta^2}, \ t \geq 0$	$a_j = \theta x_j$	A.6	A.7
Weibull	$\frac{\nu t^{\nu-1}}{\theta^\nu}e^{-(t/\theta)^\nu}, \ t \geq 0$	$a_j = \theta x_j^{1/\nu}$	A.8	A.9
Minimum extreme value	$\frac{1}{\sigma}\exp\{\frac{t-\mu}{\sigma} - \exp(\frac{t-\mu}{\sigma})\}, \ t \in \mathbf{R}$	$a_j = \mu + \sigma \ln x_j$	A.8	A.9
Maximum extreme value	$\frac{1}{\sigma}\exp\{-\frac{t-\mu}{\sigma} - \exp(-\frac{t-\mu}{\sigma})\}, \ t \in \mathbf{R}$	$a_j = \mu - \sigma \ln x_j$	A.10	A.11
Normal	$\frac{1}{\sigma\sqrt{2\pi}}e^{-(t-\mu)^2/2\sigma^2}, \ t \in \mathbf{R}$	$a_j = \mu + \sigma x_j$	A.12	A.13
Lognormal	$\frac{1}{t\sigma\sqrt{2\pi}}e^{-(\ln t-\mu)^2/2\sigma^2}, \ t \geq 0$	$a_j = e^{\mu+\sigma x_j}$	A.12	A.13
Logistic	$\frac{1}{\sigma}e^{-(t-\mu)/\sigma}/(1+e^{-(t-\mu)/\sigma})^2, \ t \in \mathbf{R}$	$a_j = \mu + \sigma x_j$	A.14	A.15

Table 1.1. *Density functions and the references to the corresponding tables of AOG*

The problem of asymptotically optimal grouping by the A- and E-optimality criteria was solved for certain distribution families, and the tables of A-optimal grouping are given in [LEM 11b].

The versions of asymptotically optimal grouping maximize the test power relative to a set of close competing hypotheses, but they do not insure the largest power against some given competing hypothesis. For the given competing hypothesis H_1, it is possible to construct the χ^2 test, which has the largest power for testing hypothesis H_0 against H_1. For example, in the case of χ^2 Pearson's test, it is possible to maximize the noncentrality parameter for the given number of intervals k:

$$\max_{a_1<a_2<....<a_{k-1}} \left(n \sum_{j=1}^{k} \frac{\left(p_j^1\left(\boldsymbol{\theta}^1\right) - p_j^0\left(\boldsymbol{\theta}^0\right)\right)^2}{p_j^0\left(\boldsymbol{\theta}^0\right)} \right), \qquad [1.38]$$

where:

$$p_j^0\left(\boldsymbol{\theta}^0\right) = \int\limits_{a_{j-1}}^{a_j} f_0\left(u; \boldsymbol{\theta}^0\right) du \quad \text{and} \quad p_j^1\left(\boldsymbol{\theta}^1\right) = \int\limits_{a_{j-1}}^{a_j} f_1\left(u; \boldsymbol{\theta}^1\right) du$$

are the probabilities to fall into j-th interval according to the hypotheses H_0 and H_1, respectively. Let us refer to this grouping method as optimal grouping.

Asymptotically optimal boundary points, corresponding to different optimality criteria, as well as the optimal points, corresponding to [1.38], are considerably different from each other. For example, the boundary points maximizing criteria [1.35]–[1.38] for the following pair of competing hypotheses are given in Table 1.2. The null hypothesis H_0 is the normal distribution with the density function

$$f(t; \boldsymbol{\theta}) = \frac{1}{\sigma\sqrt{2\pi}} \exp\left\{-\frac{(t-\mu)^2}{2\sigma^2}\right\}, \qquad [1.39]$$

and the parameters $\mu = 0$, $\sigma = 1$, and the competing hypothesis H_1 is the logistic distribution with the density function

$$f(t; \boldsymbol{\theta}) = \frac{\pi}{\theta_2\sqrt{3}} \exp\left\{-\frac{\pi(t-\theta_1)}{\theta_2\sqrt{3}}\right\} \Bigg/ \left[1 + \exp\left\{-\frac{\pi(t-\theta_1)}{\theta_2\sqrt{3}}\right\}\right]^2, \qquad [1.40]$$

and the parameters $\theta_1 = 0$, $\theta_2 = 1$.

Optimality criterion	a_1	a_2	a_3	a_4	a_5	a_6	a_7	a_8
A-optimum	-2.3758	-1.6915	-1.1047	-0.4667	0.4667	1.1047	1.6915	2.3758
D-optimum	-2.3188	-1.6218	-1.0223	-0.3828	0.3828	1.0223	1.6218	2.3188
E-optimum	-1.8638	-1.1965	-0.6805	-0.2216	0.2216	0.6805	1.1965	1.8638
Optimal grouping	-3.1616	-2.0856	-1.2676	-0.4601	0.4601	1.2676	2.0856	3.1616

Table 1.2. *Optimal boundary points for $k = 9$*

Moreover, in the case of given competing hypothesis, we can use the so-called Neyman-Pearson classes [GRE 96], for which the random variable domain is partitioned into the intervals of two types, according to the inequalities $f_0(t;\boldsymbol{\theta}) < f_1(t;\boldsymbol{\theta})$ and $f_0(t;\boldsymbol{\theta}) > f_1(t;\boldsymbol{\theta})$, where $f_0(t;\boldsymbol{\theta})$ and $f_1(t;\boldsymbol{\theta})$ are the density functions, corresponding to the competing hypotheses. For H_0 and H_1 from our example, we have the first type intervals:

$$(-\infty; -2.3747], (-0.6828; 0.6828], (2.3747; \infty),$$

and the second type intervals:

$$(-2.3747; -0.6828], (0.6828; 2.3747].$$

The asymptotically D-optimal boundary points of grouping intervals I_j and probabilities $P\{X_1 \in I_j\}, j = 1, ..., k$, given in Tables A.12 and A.13 of Appendix A, can be used for testing normality and estimating both parameters μ and σ. In Table A.12, the boundary points x_j, $j = 1, \ldots, k-1$ are listed in a form that is invariant with respect to the parameters μ and σ of the normal distribution. To calculate the statistic of Pearson [1.4], the boundaries a_j separating the intervals for specified k are found using the values of x_j, taken from the corresponding row of the table: $a_j = \hat{\sigma} x_j + \hat{\mu}$, where $\hat{\mu}$ and $\hat{\sigma}$ are the maximum likelihood estimates of the parameters derived from the given complete sample. In the last column, there are the values of relative asymptotic information:

$$A = \frac{\det \mathbf{J}}{\det \mathbf{I}},$$

where $\mathbf{J}$ and $\mathbf{I}$ are the Fisher information matrixes from grouped and complete data correspondingly. The probabilities of falling into a given interval for evaluating the statistic [1.4] are taken from the corresponding row of Table A.13.

As was shown in section 1.3, in the case of using maximum likelihood estimators $\hat{\boldsymbol{\theta}}_n$ from complete data, the distribution of Pearson's statistic under H_0 significantly differs from the χ^2 distribution, and it depends on the parameter $\boldsymbol{\theta}$. However, it is possible to use the Monte Carlo simulations to approximate the distribution

$P\{X_n^2(\hat{\boldsymbol{\theta}}_n) < x|H_0\}$. When asymptotically D-optimal grouping is used in the Pearson test for H_0, corresponding to the normal distribution [1.39], the resulting percentage points of the distributions $P\{X_n^2(\hat{\boldsymbol{\theta}}_n) < x|H_0\}$ and the approximating models of limiting distributions are shown in Table 1.3, where $\beta_{\rm III}(\theta_0, \theta_1, \theta_2, \theta_3, \theta_4)$ is the type III Beta distribution with the density function:

$$f(x) = \frac{\theta_2^{\theta_0}}{\theta_3\beta(\theta_0, \theta_1)} \frac{[(x-\theta_4)/\theta_3\,]^{\theta_0-1}[1-(x-\theta_4)/\theta_3\,]^{\theta_1-1}}{[1+(\theta_2-1)(x-\theta_4)/\theta_3\,]^{\theta_0+\theta_1}}.$$

To make a decision regarding testing the hypothesis of normality, the obtained value of statistic is compared with the critical value from the corresponding row of Table 1.3, or the p-value can be obtained using the approximation of the limiting distribution model at the same row of the table, and then compared with the specified level of significance α.

k	$1-\alpha$					Limiting distribution model
	0.85	0.9	0.95	0.975	0.99	
4	2.74	3.37	4.48	5.66	7.26	$B_{III}(1.2463, 3.8690, 4.6352, 19.20, 0.005)$
5	4.18	5	6.39	7.77	9.59	$B_{III}(1.7377, 3.8338, 5.5721, 26.00, 0.005)$
6	5.61	6.54	8.09	9.61	11.62	$B_{III}(2.1007, 4.1518, 4.1369, 26.00, 0.005)$
7	6.95	7.98	9.67	11.31	13.43	$B_{III}(2.5019, 4.6186, 3.4966, 28.00, 0.005)$
8	8.28	9.4	11.21	12.95	15.22	$B_{III}(2.9487, 5.8348, 3.1706, 34.50, 0.005)$
9	9.56	10.76	12.69	14.53	16.87	$B_{III}(3.5145, 6.3582, 3.2450, 39.00, 0.005)$
10	10.84	12.11	14.16	16.12	18.58	$B_{III}(3.9756, 6.7972, 3.0692, 41.50, 0.005)$
11	12.08	13.42	15.55	17.59	20.19	$B_{III}(4.4971, 6.9597, 3.0145, 43.00, 0.005)$
12	13.34	14.74	16.98	19.1	21.77	$B_{III}(5.1055, 7.0049, 3.1130, 45.00, 0.005)$
13	14.56	16.01	18.34	20.53	23.3	$B_{III}(5.7809, 7.0217, 3.2658, 47.00, 0.005)$
14	15.78	17.29	19.68	21.96	24.81	$B_{III}(6.6673, 6.9116, 3.5932, 49.00, 0.005)$
15	16.98	18.54	21.04	23.4	26.37	$B_{III}(7.0919, 7.2961, 3.4314.51.50, 0.005)$

Table 1.3. *Percentage points for the distribution $P\{X_n^2(\hat{\boldsymbol{\theta}}_n) < x|H_0\}$ in the case of testing normality (μ and σ are estimated) and using AOG*

As can be seen from Table A.12, when using asymptotically D-optimal grouping and $k = 15$ intervals in the grouped sample, about 95 % of the information relative to the parameter vector is preserved.

Further increase of A with the number of intervals k growth is insignificant. So, we recommend choosing the number of intervals on the basis of the condition $np_j(\boldsymbol{\theta}) \geq 5$ for each $j = 1, ..., k$. For AOG, the probabilities of falling into intervals are not equal: usually, these probabilities are minimal for the outermost intervals. It should also be noted that for small sample sizes, $n = 10 - 20$, the distributions of the statistics differ substantially from their asymptotic distributions. This condition sets an upper bound estimate on the number of intervals: $k \leq k_{\max}$, where $k_{\max}$ is taken from the condition:

$$\min_{j=1,\ldots,k} \{np_j(\boldsymbol{\theta})\} > 1.$$

The number of grouping intervals affects the power of the Pearson χ^2 test. It is absolutely unnecessary that its power against a competing distribution (hypothesis) should be maximal for $k = k_{\max}$. Let us consider the power of the Pearson and Nikulin-Rao-Robson tests when testing normality against the following competing hypotheses:

H_1 corresponds to the generalized normal distribution with the density

$$f(t;\boldsymbol{\theta}) = \frac{\theta_2}{2\theta_1\Gamma(1/\theta_2)} \exp\left\{-\left(\frac{|t-\theta_0|}{\theta_1}\right)^{\theta_2}\right\}$$

and the shape parameter $\theta_2 = 4$;

H_2 corresponds to the Laplace distribution with the density

$$f(t;\boldsymbol{\theta}) = \frac{1}{2\theta_1} \exp\left\{-\frac{|t-\theta_0|}{\theta_1}\right\};$$

H_3 corresponds to the logistic distribution with the density [1.40], which is very close to the normal distribution.

Figure 1.2 shows the densities of the distributions, corresponding to hypotheses H_1, H_2 and H_3. This choice of hypotheses has a certain justification. Hypothesis H_2, corresponding to the Laplace distribution, is the most distant from H_0. Usually, there are no problems in distinguishing them. The logistic distribution (hypothesis H_3) is very close to the normal law and it is generally difficult to distinguish them by goodness-of-fit tests. The competing hypothesis H_1, which corresponds to the generalized normal distribution with a shape factor $\theta_2 = 4$, is a "litmus test" for detection of hidden deficiencies in some tests: as was shown by Lemeshko and Lemeshko in [LEM 05], for small sample sizes n and small probabilities α of type I error, some special normality tests are not able to distinguish close distributions from normal. In these cases, the power $1 - \beta$ with respect to hypothesis H_1, where β is the probability of a type II error, is smaller than α. This means that the distribution corresponding to H_1 is "more normal" than the normal law and indicates that such tests are biased.

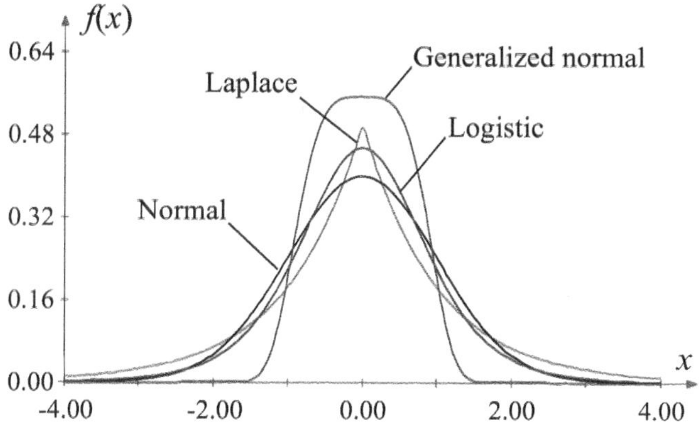

Figure 1.2. *Probability density functions corresponding to the considered hypotheses H_i. For a color version of this figure, see www.iste.co.uk/nikulin/chisquared.zip*

The power of the Pearson χ^2 test was studied with different numbers of intervals $k \leq k_{\max}$ for specified sample sizes n. In Table 1.4, there are the estimates of power of Pearson's test relative to the competing hypotheses H_1, H_2 and H_3, and corresponding to the optimal number k_{opt} of grouping intervals. To a certain extent, it is possible to orient ourselves in choosing k on the basis of the values of k_{opt} as a function of n listed in Table 1.4.

n	$k_{\max}$	k_{opt}	α				
			0.15	0.1	0.05	0.025	0.01
H_1							
10	4	4	0.235	0.146	0.043	0.032	0.002
20	4	5	0.262	0.177	0.100	0.058	0.021
30	5	5	0.312	0.216	0.136	0.079	0.043
40	6	5	0.336	0.267	0.168	0.111	0.061
50	6	5	0.401	0.311	0.204	0.129	0.068
100	9	5	0.558	0.479	0.352	0.254	0.158
150	10	7	0.722	0.634	0.486	0.353	0.217
200	11	9	0.783	0.695	0.548	0.417	0.279
300	13	11	0.907	0.858	0.756	0.646	0.492
H_2							
10	4	4	0.267	0.206	0.074	0.058	0.01
20	4	4	**0.264**	0.177	0.104	**0.067**	**0.037**
		5	0.247	**0.189**	**0.116**	0.061	0.024
30	5	5	0.312	0.261	0.153	0.103	0.044
40	6	7	0.443	0.358	0.25	0.167	0.101
50	6	7	0.5	0.423	0.312	0.225	0.138
100	9	9	0.77	0.708	0.596	0.494	0.379
150	10	9	0.899	0.86	0.785	0.705	0.596
200	11	11	0.964	0.946	0.908	0.88	0.786
300	13	13	0.996	0.993	0.985	0.974	0.95
H_3							
10	4	4	0.221	0.15	0.046	0.034	0.003
20	4	4	0.194	0.125	0.059	0.038	0.016
30	5	5	0.169	0.125	0.062	0.034	0.012
40	6	7	0.204	0.143	0.082	0.045	0.02
50	6	7	0.214	0.155	0.088	0.05	0.023
100	9	10	0.303	0.231	0.146	0.09	0.047
150	10	10	0.359	0.284	0.191	0.124	0.072
200	11	11	0.432	0.355	0.25	0.175	0.105
300	13	13	0.566	0.486	0.373	0.28	0.19

Table 1.4. *Estimates of the power of Pearson's χ^2 test with respect to hypotheses H_1, H_2, and H_3*

Estimates of the power of the Nikulin-Rao-Robson test for the competing hypotheses H_1, H_2 and H_3 for k_{opt} are given in Table 1.5. This test is generally more powerful than the Pearson test (for example, see its powers relative to the competing hypotheses H_2 and H_3). Here, we often have $k_{opt} = k_{\max}$ for

$$\min_{j=1,\dots,k} \{np_j(\boldsymbol{\theta})\} > 1.$$

n	$k_{\max}$	k_{opt}	α				
			0.15	0.1	0.05	0.025	0.01
			H_1				
10	4	4	0.348	0.1	0.029	0.009	0.006
20	4	5	0.234	0.143	0.074	0.041	0.016
30	5	5	0.256	0.197	0.102	0.053	0.023
40	6	5	0.293	0.221	0.123	0.079	0.035
50	6	5	0.326	0.240	0.148	0.083	0.040
100	9	5	0.485	0.395	0.271	0.179	0.102
150	10	6	0.619	0.530	0.397	0.284	0.179
		7	0.641	0.539	0.383	0.261	0.148
200	11	9	0.713	0.616	0.464	0.339	0.214
300	13	11	0.872	0.810	0.695	0.573	0.420
			H_2				
10	4	4	0.368	0.103	0.055	0.031	0.007
20	4	5	0.250	0.210	0.126	0.065	0.039
30	5	6	0.349	0.265	0.185	0.127	0.078
40	6	7	0.474	0.403	0.297	0.218	0.149
50	6	7	0.548	0.473	0.365	0.281	0.190
100	9	9	0.807	0.755	0.667	0.583	0.482
150	10	9	0.919	0.889	0.834	0.774	0.691
200	11	11	0.973	0.961	0.933	0.900	0.849
300	13	11	0.997	0.995	0.990	0.983	0.968
		13	0.997	0.995	0.990	0.983	0.968
			H_3				
10	4	4	0.321	0.083	0.034	0.014	0.005
20	4	5	0.166	0.120	0.065	0.030	0.014
30	5	6	0.198	0.138	0.080	0.047	0.024
40	6	7	0.232	0.173	0.104	0.063	0.034
50	6	7	0.251	0.188	0.117	0.074	0.040
100	9	10	0.360	0.290	0.202	0.141	0.091
150	10	10	0.432	0.358	0.263	0.195	0.131
200	11	11	0.509	0.436	0.337	0.259	0.183
300	13	13	0.641	0.572	0.469	0.381	0.288

Table 1.5. *Estimates of the power of the Nikulin–Rao–Robson χ^2 test with respect to hypotheses H_1, H_2, and H_3*

However, this is not always so. In terms of its power relative to the "tricky" hypothesis H_1, k_{opt} is considerably smaller than $k_{\max}$ with AOG.

So, the power of the Pearson and Nikulin-Rao-Robson tests can be maximized by the optimal selection of the number of intervals and interval boundary points.

2

Chi-squared Test for Censored Data

In reliability or survival studies, the observed data are usually of the form:

$$\mathbf{X} = (X_1, \delta_1), (X_2, \delta_2), ..., (X_n, \delta_n),$$

where $X_i = \min(T_i, C_i)$ is the observation value, T_i is the lifetime and C_i is a censoring time. The censoring indicator $\delta_i = I\{T_i \le C_i\}$.

There are various forms of right censoring (for details, see, for example, [KLE 03], [AND 93], [BAG 13], [BAG 11], [SAS 08] and [KIM 93]):

– If individuals are observed up to some predetermined time c, then censoring is referred to as Type I censoring and in the case when $\delta_i = 0 : X_i = c$.

– If a life test is terminated whenever a specified number of failures (say, m) has occurred, it is referred to as Type II censoring and in this case when $\delta_i = 0 : X_i = X_{(m)}$, where $X_{(m)}$ is the last observed failure time.

– Let the lifetime T and the censoring time C be independent random variables from the distribution functions $F(t)$ and $F^C(t)$, respectively. All lifetimes and censoring times are assumed mutually independent. Then, this form of censoring is referred to as independent random censoring.

We consider the composite hypothesis:

$$H_0 : F(t) \in \{F(t;\boldsymbol{\theta}), \boldsymbol{\theta} \in \Theta \subset \mathbf{R}^s\},$$

meaning that the cumulative distribution function of lifetimes $T_1, ..., T_n$ belongs to the class of distributions of the form $F(t;\boldsymbol{\theta})$, where $\boldsymbol{\theta} = (\theta_1, ..., \theta_s)^T \in \Theta \subset \mathbf{R}^s$ is an unknown $s-$dimensional parameter. The probability distribution can also be defined by the *survival function*

$$S(t;\boldsymbol{\theta}) = 1 - F(t;\boldsymbol{\theta}) = \mathbf{P}\{T_1 > t\}$$

and the *hazard function*

$$\lambda(t;\boldsymbol{\theta}) = \lim_{h\downarrow 0} \frac{1}{h}\mathbf{P}_{\boldsymbol{\theta}}\{t < T_1 \leq t + h | T_1 > t\} = \frac{f(t;\boldsymbol{\theta})}{S(t;\boldsymbol{\theta})}.$$

For any $t > 0$, the value $S(t;\boldsymbol{\theta})$ of the survival function is the probability of not failing to time t, and the value $\lambda(t;\boldsymbol{\theta})$ of the hazard rate characterizes the failure risk just after the time t for the objects that survive to this time.

2.1. Generalized Pearson-Fisher chi-squared test

Chi-square type tests require dividing an observed interval $[0, \tau]$ into k smaller intervals $I_j = (a_{j-1}, a_j]$, $0 = a_0 < a_1 < \ldots < a_{k-1} < a_k = \tau$.

Let $\hat{F}_n(t)$ be the Kaplan-Meier estimator [KAP 58] of the lifetime distribution F. Define the empirical probabilities to fall into grouping intervals as:

$$\hat{p}_j = \hat{F}_n(a_j) - \hat{F}_n(a_{j-1}), j = 1, ..., k,$$

where $\hat{p}_1$ and $\hat{p}_k$ must be defined as $\hat{p}_1 = \hat{F}_n(a_1)$ and $\hat{p}_k = 1 - \hat{F}_n(a_{k-1})$. Denote by

$$p_j(\boldsymbol{\theta}) = F(a_j;\boldsymbol{\theta}) - F(a_{j-1};\boldsymbol{\theta}),$$

the theoretical probabilities to fall into grouping intervals. Unknown parameters of the tested distribution are estimated as follows:

$$\tilde{\boldsymbol{\theta}}_n = Arg \min_{\boldsymbol{\theta}} \xi_n^T(\boldsymbol{\theta})\xi_n(\boldsymbol{\theta}),$$

where

$$\xi_n(\boldsymbol{\theta}) = \left(\frac{n\hat{p}_1 - np_1(\boldsymbol{\theta})}{\sqrt{np_1(\boldsymbol{\theta})}}, ..., \frac{n\hat{p}_k - np_k(\boldsymbol{\theta})}{\sqrt{np_k(\boldsymbol{\theta})}} \right)^T.$$

It is well known [LI 93] that $\xi_n(\tilde{\boldsymbol{\theta}}_n) \xrightarrow{d} N(0, \Sigma)$, where

$$\Sigma = PDJ\Sigma^{(1)}J^T DP,$$

$$P = I - C\left(C^T C\right)^{-1} C^T, \quad C = D\left(\frac{\partial}{\partial\boldsymbol{\theta}} p(\boldsymbol{\theta})\right)^T,$$

$$p(\boldsymbol{\theta}) = (p_1(\boldsymbol{\theta}), ..., p_k(\boldsymbol{\theta}))^T,$$

$$D = \text{diag}\left((p_1(\boldsymbol{\theta}))^{-1/2}, ..., (p_k(\boldsymbol{\theta}))^{-1/2}\right),$$

$$J = \begin{pmatrix} 1 & 0 & 0 & \dots & 0 \\ -1 & 1 & 0 & \dots & 0 \\ 0 & -1 & 1 & \dots & 0 \\ \vdots & \vdots & \vdots & \ddots & \vdots \\ 0 & 0 & 0 & \dots & 1 \\ 0 & 0 & 0 & \dots & -1 \end{pmatrix}_{k\times(k-1)}.$$

The elements of the $(k-1) \times (k-1)$ matrix $\Sigma^{(1)}$ are calculated by the formula:

$$\sigma_{ij}^{(1)} = (1 - F(a_i; \boldsymbol{\theta})) \cdot (1 - F(a_j; \boldsymbol{\theta})) \cdot v\left(\min(a_i, a_j)\right),$$

where the variance function:

$$v(t) = \int_0^t \frac{dF(u; \boldsymbol{\theta})}{(1 - F(u; \boldsymbol{\theta}))^2 (1 - F^C(u))}.$$

Here, the Kaplan-Meier estimator, obtained by the inverted sample, in which δ_i is replaced by $1 - \delta_i$, can be used as the distribution of censoring times $\hat{F}^C(t)$. Let $\hat{\Sigma}$ be obtained by using the minimum chi-square estimator $\tilde{\boldsymbol{\theta}}_n$ as $\boldsymbol{\theta}$ in formulas for the matrix Σ, and $\hat{\Sigma}^+$ denotes the Moore-Penrose inverse of $\hat{\Sigma}$. Then, under H_0, the statistic of the generalized Pearson-Fisher test

$$X_n^2 = \xi_n^T\left(\tilde{\boldsymbol{\theta}}_n\right)\hat{\Sigma}^+\xi_n\left(\tilde{\boldsymbol{\theta}}_n\right)$$

has the χ^2-distribution with $k - s - 1$ degrees of freedom as $n \to \infty$ [LI 93].

2.2. Maximum likelihood estimators for censored data

Suppose that the lifetimes $T_1, \cdots, T_n$ are absolutely continuous i.i.d. random variables, the p.d.f. of the random variable T_1 belongs to a parametric family $\{f(\cdot;\boldsymbol{\theta}),\, \boldsymbol{\theta} \in \boldsymbol{\Theta} \subset \mathbf{R}^s\}$, where $f(\cdot;\boldsymbol{\theta})$ is the density of T_i with respect to σ-finite measure μ and $\boldsymbol{\Theta}$ is an open.

We denote by $\bar{G}_i$ the survival function of the censoring time C_i. For any $t > 0$, the value $\bar{G}_i(t)$ is probability for the i-th object not be censored to time t.

Let us consider the distribution of the random vector (X_i, δ_i) in the case of random censoring with absolutely continuous censoring times C_i with the p.d.f. $g_i(t)$.

In this case, we obtain that the following expressions for the likelihood function is:

$$L(\boldsymbol{\theta}) = \prod_{i=1}^{n} f^{\delta_i}(X_i;\boldsymbol{\theta})S^{1-\delta_i}(X_i;\boldsymbol{\theta})\,\bar{G}_i^{\delta_i}(X_i)\,g_i^{1-\delta_i}(X_i), \quad \boldsymbol{\theta} \in \boldsymbol{\Theta}. \qquad [2.1]$$

DEFINITION 2.1.– *Right censoring is called non-informative if the survival functions* $\bar{G}_i$ *do not depend on the parameter* $\boldsymbol{\theta} = (\theta_1, ..., \theta_s)^T$.

Since the problem is to estimate the parameter $\boldsymbol{\theta}$, we can skip the multipliers which do not depend on this parameter. So under non-informative censoring, the likelihood function has the next form:

$$L(\boldsymbol{\theta}) = \prod_{i=1}^{n} f^{\delta_i}(X_i;\boldsymbol{\theta}) S^{1-\delta_i}(X_i;\boldsymbol{\theta}), \quad \boldsymbol{\theta} \in \boldsymbol{\Theta}. \qquad [2.2]$$

Using the relation $f(t;\boldsymbol{\theta}) = \lambda(t;\boldsymbol{\theta})S(t;\boldsymbol{\theta})$, the likelihood function can be written as:

$$L(\boldsymbol{\theta}) = \prod_{i=1}^{n} \lambda^{\delta_i}(X_i;\boldsymbol{\theta}) S(X_i;\boldsymbol{\theta}), \quad \boldsymbol{\theta} \in \boldsymbol{\Theta}. \qquad [2.3]$$

The estimator $\hat{\boldsymbol{\theta}}$, maximizing the likelihood function $L(\boldsymbol{\theta}), \boldsymbol{\theta} \in \boldsymbol{\Theta}$, is called the *maximum likelihood estimator*. The log-likelihood function

$$\ell(\boldsymbol{\theta}) = \sum_{i=1}^{n} \delta_i \ln \lambda(X_i;\boldsymbol{\theta}) + \sum_{i=1}^{n} \ln S(X_i;\boldsymbol{\theta}), \quad \boldsymbol{\theta} \in \boldsymbol{\Theta}, \qquad [2.4]$$

is maximized at the same point as the likelihood function.

If $\lambda(u;\boldsymbol{\theta})$ is a sufficiently smooth function of the parameter $\boldsymbol{\theta}$, then the ML estimator satisfies the equation:

$$\dot{\ell}(\hat{\boldsymbol{\theta}}) = \mathbf{0}; \qquad [2.5]$$

here $\mathbf{0} = \mathbf{0}_s = (0, ..., 0)^T \in R^s$, and $\dot{\ell}$ is the score vector:

$$\dot{\ell}(\boldsymbol{\theta}) = \frac{\partial}{\partial \boldsymbol{\theta}} \ell(\boldsymbol{\theta}) = \left(\frac{\partial}{\partial \theta_1} \ell(\boldsymbol{\theta}), \cdots, \frac{\partial}{\partial \theta_s} \ell(\boldsymbol{\theta}) \right)^T. \qquad [2.6]$$

Equalities [2.5] imply:

$$\dot{\ell}(\boldsymbol{\theta}) = \sum_{i=1}^{n} \delta_i \frac{\partial}{\partial \boldsymbol{\theta}} \ln \lambda(X_i;\boldsymbol{\theta}) + \sum_{i=1}^{n} \frac{\partial}{\partial \boldsymbol{\theta}} \ln S(X_i;\boldsymbol{\theta})$$
$$= \sum_{i=1}^{n} \delta_i \frac{\partial}{\partial \boldsymbol{\theta}} \ln \lambda(X_i;\boldsymbol{\theta}) - \sum_{i=1}^{n} \frac{\partial}{\partial \boldsymbol{\theta}} \Lambda(X_i;\boldsymbol{\theta}),$$

where

$$\Lambda(t;\boldsymbol{\theta}) = -\ln S(t;\boldsymbol{\theta}) = \int_0^t \lambda(u;\boldsymbol{\theta})du.$$

Fisher's information matrix is:

$$\boldsymbol{I}(\boldsymbol{\theta}) = -\mathbf{E}_{\boldsymbol{\theta}}\ddot{\ell}(\boldsymbol{\theta}), \qquad [2.7]$$

where

$$\ddot{\ell}(\boldsymbol{\theta}) = \sum_{i=1}^{n} \delta_i \frac{\partial^2}{\partial \boldsymbol{\theta}^2} \ln \lambda(X_i;\boldsymbol{\theta}) - \sum_{i=1}^{n} \frac{\partial^2}{\partial \boldsymbol{\theta}^2} \Lambda(X_i;\boldsymbol{\theta}). \qquad [2.8]$$

In the case of non-informative and random censoring, the asymptotic properties of ML estimators result from general results of asymptotic statistics (see, for example, [VAN 98]). Let us denote by $\boldsymbol{\theta}_0$ the true value of $\boldsymbol{\theta}$. In this case under some regularity conditions, we may write that:

$$\hat{\boldsymbol{\theta}} \xrightarrow{P} \boldsymbol{\theta}_0,\ \sqrt{n}(\hat{\boldsymbol{\theta}}-\boldsymbol{\theta}_0) = \boldsymbol{i}^{-1}(\boldsymbol{\theta}_0)\frac{1}{\sqrt{n}}\dot{\ell}(\boldsymbol{\theta}_0)+o_P(\mathbf{1}),\ -\frac{1}{n}\ddot{\ell}(\hat{\boldsymbol{\theta}}) =\xrightarrow{P} \boldsymbol{i}(\boldsymbol{\theta}_0), \qquad [2.9]$$

$$\sqrt{n}(\hat{\boldsymbol{\theta}} - \boldsymbol{\theta}_0) \xrightarrow{d} N_s(\mathbf{0}, \boldsymbol{i}^{-1}(\boldsymbol{\theta}_0)), \qquad [2.10]$$

$$\frac{1}{\sqrt{n}}\dot{\ell}(\boldsymbol{\theta}_0) \xrightarrow{d} N_s(\mathbf{0}, \boldsymbol{i}(\boldsymbol{\theta}_0)), \qquad [2.11]$$

where $\hat{\boldsymbol{\theta}}$ is the root of equation [2.5] and the matrix $\boldsymbol{i}(\boldsymbol{\theta}_0) = \lim_{n\to\infty} \boldsymbol{I}(\boldsymbol{\theta}_0)/n$. Let us write the log-likelihood function $\ell(\boldsymbol{\theta})$ and the score vector $\dot{\ell}(\boldsymbol{\theta})$ for the observed data in terms of stochastic processes $N_i(t)$ and $Y_i(t)$.

The trajectories of the counting processes N_i have the form:

$$N_i(t) = \begin{cases} 0, \ 0 \le t < X_i, \\ 1, \quad t \ge X_i, \end{cases} \qquad [2.12]$$

if $\delta_i = 1$, and $N_i(t) \equiv 0$, if $\delta_i = 0$. So:

$$\int_0^\infty \ln \lambda(u;\boldsymbol{\theta}) dN_i(u) = \begin{cases} \ln \lambda(X_i;\boldsymbol{\theta}), & \delta_i = 1, \\ 0, & \delta_i = 0. \end{cases} = \delta_i \ln \lambda(X_i;\boldsymbol{\theta}). \quad [2.13]$$

The trajectories of the stochastic process Y_i have the form:

$$Y_i(t) = \begin{cases} 1, \; 0 \le t \le X_i, \\ 0, \quad t > X_i. \end{cases} \quad [2.14]$$

So:

$$\int_0^\infty Y_i(u)\lambda(u;\boldsymbol{\theta})du = \int_0^{X_i} \lambda(u;\boldsymbol{\theta})du = -\ln S(X_i;\boldsymbol{\theta}), \quad \{\boldsymbol{\theta} \in \boldsymbol{\Theta}\}. \quad [2.15]$$

These relations imply that under non-informative and random censoring, the considered functions may be written in the form:

$$\ell(\boldsymbol{\theta}) = \sum_{i=1}^n \int_0^\infty \{\ln \lambda(u;\boldsymbol{\theta})dN_i(u) - Y_i(u)\lambda(u;\boldsymbol{\theta})\}du, \quad \{\boldsymbol{\theta} \in \boldsymbol{\Theta}\}, \quad [2.16]$$

from where it follows that:

$$\begin{aligned} \dot{\ell}(\boldsymbol{\theta}) &= \sum_{i=1}^n \int_0^\infty \frac{\partial}{\partial \boldsymbol{\theta}} \ln \lambda(u;\boldsymbol{\theta}) \{dN_i(u) - Y_i(u)\lambda(u;\boldsymbol{\theta})du\} \\ &= \sum_{i=1}^n \int_0^\infty \frac{\partial}{\partial \boldsymbol{\theta}} \ln \lambda(u;\boldsymbol{\theta}) \, dM_i(u;\boldsymbol{\theta}), \quad \{\boldsymbol{\theta} \in \boldsymbol{\Theta}\} \end{aligned}$$

and

$$\ddot{\ell}(\boldsymbol{\theta}) = \sum_{i=1}^n \int_0^\infty \frac{\partial^2}{\partial \boldsymbol{\theta}^2} \ln \lambda(u;\boldsymbol{\theta}) \, dM_i(u;\boldsymbol{\theta})$$

$$-\sum_{i=1}^{n}\int_0^{\infty}\frac{\partial}{\partial\boldsymbol{\theta}}\ln\lambda(u;\boldsymbol{\theta})\,(\frac{\partial}{\partial\boldsymbol{\theta}}\ln\lambda(u;\boldsymbol{\theta}))^T\lambda(u;\boldsymbol{\theta})Y_i(u)du,\quad\{\boldsymbol{\theta}\in\boldsymbol{\Theta}\},$$

where

$$M_i(t;\boldsymbol{\theta}) = N_i(t) - A_i(t) = N_i(t) - \int_0^t Y_i(u)\lambda(u;\boldsymbol{\theta})du,\quad\{\boldsymbol{\theta}\in\boldsymbol{\Theta}\},\tag{2.17}$$

is martingale of counting process $N_i(t), t \geq 0$. Let us consider the following processes:

$$N(t) = \sum_{i=1}^{n} N_i(t)\quad\text{and}\quad Y(t) = \sum_{i=1}^{n} Y_i(t),\quad t \geq 0.\tag{2.18}$$

By tradition, accepted in survival analysis and reliability, we suppose that the processes N_i and Y_i are observed during the finite time $\tau > 0$. It means that at time τ, observation of all objects is censored, so in the place of censoring time C_i, censoring time $C_i \wedge \tau$ is used. We denote them once more by C_i. The process $N(t)$ shows, for any $t > 0$, the number of observed failures in the interval $[0,\tau]$ and the process $Y(t)$ shows the number of objects which are *at risk* (not failed, not truncated and not censored) just prior to the time $t, t < \tau$.

We note that in our case the Fisher information matrix can be written as:

$$\boldsymbol{I}(\boldsymbol{\theta}) = -\mathbf{E}_{\boldsymbol{\theta}}\ddot{\ell}(\boldsymbol{\theta}) = \mathbf{E}_{\boldsymbol{\theta}}\sum_{i=1}^{n}\int_0^{\tau}\frac{\partial}{\partial\boldsymbol{\theta}}\ln\lambda(u;\boldsymbol{\theta})$$
$$\left(\frac{\partial}{\partial\boldsymbol{\theta}}\ln\lambda(u;\boldsymbol{\theta})\right)^T\lambda(u;\boldsymbol{\theta})Y_i(u)du.\tag{2.19}$$

2.3. Nikulin-Rao-Robson chi-squared test for censored data

Divide the interval $[0,\tau]$ into $k > s$ smaller intervals $I_j = (a_{j-1},\ a_j]$. $a_0 = 0$, $a_k = \tau$ and denote by

$$U_j = N(a_j) - N(a_{j-1})\tag{2.20}$$

the number of observed failures in the j-th interval I_j, $j = 1, 2, ..., k$. What is the "expected" number of observed failures in the interval I_j under the hypothesis H_0? Taking into account the equality

$$\mathbf{E}N(t) = \mathbf{E}\int_0^t \lambda(u; \boldsymbol{\theta}_0) Y(u) du, \quad [2.21]$$

we can "expect" to observe

$$E_j = \int_{a_{j-1}}^{a_j} \lambda(t; \hat{\boldsymbol{\theta}}) Y(u) du \quad [2.22]$$

failures, where $\hat{\boldsymbol{\theta}}$ is the MLE of the parameter $\boldsymbol{\theta}$, given by [2.5], and $\lambda(t; \boldsymbol{\theta})$ is the hazard function. The chi-squared test, considered by Akritas in [AKR 88], is based on the vector:

$$\boldsymbol{Z} = (Z_1, ..., Z_k)^T, \quad Z_j = \frac{1}{\sqrt{n}}(U_j - E_j), \quad j = 1, ..., k. \quad [2.23]$$

To investigate the properties of the statistic $\boldsymbol{Z}$, we need the properties of the stochastic process

$$H_n(t) = \frac{1}{\sqrt{n}}(N(t) - \int_0^t \lambda(u; \hat{\boldsymbol{\theta}}) Y(u) du) \quad [2.24]$$

in the interval $[0, \tau]$. We recall that consistency and asymptotic normality of the ML estimator $\hat{\boldsymbol{\theta}}$ hold under the following sufficient conditions. **Conditions A:**

1) There exists a neighborhood Θ_0 of $\boldsymbol{\theta}_0$ such that for all n and $\boldsymbol{\theta} \in \Theta_0$, and almost all $t \in [0, \tau]$, the partial derivatives of $\lambda(t; \boldsymbol{\theta})$ of the first, second, and the third order with respect to $\boldsymbol{\theta}$ exist and are continuous in $\boldsymbol{\theta}$ for $\boldsymbol{\theta} \in \Theta_0$. Moreover, they are bounded in $[0, \tau] \times \Theta_0$ and the log-likelihood function [2.4] may be differentiated three times with respect to $\boldsymbol{\theta} \in \Theta_0$ by interchanging the order of integration and differentiation.

2) $\lambda(t; \boldsymbol{\theta})$ is bounded away from zero in $[0, \tau] \times \Theta_0$.

3) There exists a positive deterministic function $y(t)$ such that:

$$\sup_{t\in[0,\tau]} |Y(t)/n - y(t)| \xrightarrow{P} 0.$$

4) The matrix $\boldsymbol{i}(\boldsymbol{\theta}_0) = \lim_{n\to\infty} \boldsymbol{I}(\boldsymbol{\theta}_0)/n$ (the limit exists under the conditions 1–3) is positively definite.

LEMMA 2.1.– *Under the conditions A, the following convergence holds:*

$$H_n \xrightarrow{d} V \quad \text{on} \quad D[0,\tau],$$

where V *is a zero-mean Gaussian martingale such that, for all* $0 \le s \le t$,

$$\mathbf{cov} V(s), V(t) = A(s) - \boldsymbol{C}^T(s)\boldsymbol{i}^{-1}(\boldsymbol{\theta}_0)\boldsymbol{C}(t), \qquad [2.25]$$

with

$$A(t) = \int_0^t \lambda(u;\boldsymbol{\theta}_0)y(u)du,$$

$$\boldsymbol{C}(t) = \int_0^t \frac{\partial}{\partial\boldsymbol{\theta}} \ln\lambda(u;\boldsymbol{\theta}_0)\lambda(u;\boldsymbol{\theta}_0)y(u)du = (C_1(t), ..., C_m(t))^T,$$

and $\xrightarrow{d}$ *means week convergence in the space* $D[0,\tau]$ *of cadlag functions with Skorokhod metric.*

PROOF.– Taking into account that (we use consistence of $\hat{\boldsymbol{\theta}}$ and the Taylor formula):

$$\sqrt{n}(\lambda(u;\boldsymbol{\theta}_0) - \lambda(u;\hat{\boldsymbol{\theta}})) = -\dot{\lambda}^T(u;\boldsymbol{\theta}_0)\,\sqrt{n}(\hat{\boldsymbol{\theta}} - \boldsymbol{\theta}_0) + o_P(1),$$

where $o_P(1)$ tends to zero in probability uniformly on $[0,\tau]$, and that by the property (2.10) of the ML estimator:

$$\sqrt{n}(\hat{\boldsymbol{\theta}} - \boldsymbol{\theta}_0) = \boldsymbol{i}^{-1}(\boldsymbol{\theta}_0)\,\frac{1}{\sqrt{n}}\dot{\ell}(\boldsymbol{\theta}_0) + o_P(1)$$
$$= \boldsymbol{i}^{-1}(\boldsymbol{\theta}_0)\,\frac{1}{\sqrt{n}}\int_0^\tau \frac{\partial}{\partial\boldsymbol{\theta}} \ln\lambda(u;\boldsymbol{\theta}_0)\,dM(u) + o_P(1),$$

we have:

$$H_n(t) = \frac{1}{\sqrt{n}}(N(t) - \int_0^t \lambda(u;\boldsymbol{\theta}_0)Y(u)du)$$

$$+\frac{1}{\sqrt{n}}\int_0^t (\lambda(u;\boldsymbol{\theta}_0) - \lambda(u;\hat{\boldsymbol{\theta}}))Y(u)du$$

$$= \frac{1}{\sqrt{n}}M(t) - \frac{1}{n}\int_0^t \dot{\lambda}^T(u;\boldsymbol{\theta}_0)Y(u)du\,\sqrt{n}(\hat{\boldsymbol{\theta}} - \boldsymbol{\theta}_0) + o_P(1)$$

$$= \frac{1}{\sqrt{n}}M(t) - \boldsymbol{C}^T(t)\,\boldsymbol{i}^{-1}(\boldsymbol{\theta}_0)\frac{1}{\sqrt{n}}\int_0^\tau \frac{\partial}{\partial\boldsymbol{\theta}}\ln\lambda(u;\boldsymbol{\theta}_0)\,dM(u) + o_P(1)$$

$$=: M_1(t) - \boldsymbol{C}^T(t)\,\boldsymbol{i}^{-1}(\boldsymbol{\theta}_0)M_2.$$

Conditions A imply the following convergence in probability of the predictable variations and covariations of the martingales M_1 and M_2:

$$< M_1 > (t) = \frac{1}{n}\int_0^t \lambda(u;\boldsymbol{\theta}_0)Y(u)du \xrightarrow{P} A(t),$$

$$< M_1, M_2 > (t) = \frac{1}{n}\int_0^t \dot{\lambda}(u;\boldsymbol{\theta}_0)Y(u)du \xrightarrow{P} C(t),$$

$$< M_2 > = \frac{1}{n}\int_0^\tau \frac{\partial}{\partial\boldsymbol{\theta}}\ln\lambda(u;\boldsymbol{\theta}_0)$$
$$\left(\frac{\partial}{\partial\boldsymbol{\theta}}\ln\lambda(u;\boldsymbol{\theta}_0)\right)^T \lambda(u;\boldsymbol{\theta}_0)Y(u)du \xrightarrow{P} \boldsymbol{i}(\boldsymbol{\theta}_0).$$

The result of the lemma is implied by the CLT for martingales because the Lindeberg condition is obviously satisfied: for all $t \in [0,\tau]$:

$$| < \frac{1}{\sqrt{n}}\int_0^t \mathbf{1}_{\{\frac{1}{\sqrt{n}} > \varepsilon\}}\, dM(u) > |$$

$$= \frac{1}{n}\mathbf{1}_{\{\frac{1}{\sqrt{n}} > \varepsilon\}}\int_0^t \lambda(u;\boldsymbol{\theta}_0)Y(u)du \le \mathbf{1}_{\{\frac{1}{\sqrt{n}} > \varepsilon\}} c_0\tau \to 0,$$

$$|<\frac{1}{\sqrt{n}}\int_0^\tau \frac{\partial}{\partial\theta_i}\ln\lambda(u;\boldsymbol{\theta}_0)\mathbf{1}_{\{\frac{1}{\sqrt{n}}\frac{\partial}{\partial\theta_i}\ln\lambda(u;\theta_0)>\varepsilon\}}\,dM(u)>|$$

$$=\frac{1}{n}\int_0^\tau\left(\frac{\partial}{\partial\theta_i}\lambda(u;\boldsymbol{\theta}_0)\right)^2\mathbf{1}_{\{\frac{1}{\sqrt{n}}\frac{\partial}{\partial\theta_i}\ln\lambda(u;\theta_0)>\varepsilon\}}\lambda(u;\boldsymbol{\theta}_0)Y(u)du$$

$$\leq d_i\tau\mathbf{1}_{\{\frac{c_i}{\sqrt{n}}>\varepsilon\}}\to 0;$$

here c_i and d_i are positive constants. The proof is complete.

Set for $i=1,...,s$ and $j,j'=1,...,k$

$$V_j=V(a_j)-V(a_{j-1}),\quad v_{jj'}=\mathbf{cov}(V_j,V_{j'}),$$

$$A_j=A(a_j)-A(a_{j-1}),$$

$$\boldsymbol{C}_j=(C_{1j},...,C_{sj})^T=\boldsymbol{C}(a_j)-\boldsymbol{C}(a_{j-1}),$$

$$\boldsymbol{V}=[v_{jj'}]_{k\times k},\quad \boldsymbol{C}=[C_{ij}]_{s\times k},$$

and denote by $\boldsymbol{A}$ the $k\times k$ diagonal matrix with diagonal elements $A_1,...,A_k$.

THEOREM 2.1.– *Under conditions A*

$$\boldsymbol{Z}\overset{d}{\to}\boldsymbol{Y}\sim N_k(\boldsymbol{0},\boldsymbol{V}),\quad \textit{as } n\to\infty, \qquad [2.26]$$

where

$$\boldsymbol{V}=\boldsymbol{A}-\boldsymbol{C}^T\boldsymbol{i}^{-1}(\boldsymbol{\theta}_0)\boldsymbol{C}. \qquad [2.27]$$

REMARK 2.1.– Set

$$\boldsymbol{G}=\boldsymbol{i}-\boldsymbol{C}\boldsymbol{A}^{-1}\boldsymbol{C}^T. \qquad [2.28]$$

Using the well-known formula for the general inverse $\mathbf{V}^-$ of the matrix $\boldsymbol{V}$ of the form [2.27], we may write:

$$\boldsymbol{V}^-=\boldsymbol{A}^{-1}+\boldsymbol{A}^{-1}\boldsymbol{C}^T\boldsymbol{G}^-\boldsymbol{C}\boldsymbol{A}^{-1}. \qquad [2.29]$$

We see that we need to invert only the diagonal $k\times k$ matrix $\boldsymbol{A}$ and find the general inverse of the $s\times s$ matrix $\boldsymbol{G}=\boldsymbol{i}-\boldsymbol{C}\boldsymbol{A}^{-1}\boldsymbol{C}^T$ (usually $s=1$, $s=2$ or $s=3$).

THEOREM 2.2.– *Under conditions A, the following estimators of* A_j, $\boldsymbol{C}_j$, $\boldsymbol{i}(\boldsymbol{\theta}_0)$ *and* $\boldsymbol{V}$ *are consistent:*

$$\hat{A}_j = \frac{U_j}{n}, \quad \hat{\boldsymbol{C}}_j = \frac{1}{n}\int_{a_{j-1}}^{a_j} \frac{\partial}{\partial \boldsymbol{\theta}} \ln \lambda(u; \hat{\boldsymbol{\theta}}) dN(u),$$

and

$$\hat{\boldsymbol{i}} = \frac{1}{n}\int_0^{\tau} \frac{\partial}{\partial \boldsymbol{\theta}} \ln \lambda(u; \hat{\boldsymbol{\theta}}) \left(\frac{\partial}{\partial \boldsymbol{\theta}} \ln \lambda(u; \hat{\boldsymbol{\theta}})\right)^T dN(u),$$

$$\hat{\boldsymbol{V}} = \hat{\boldsymbol{A}} - \hat{\boldsymbol{C}}^T \hat{\boldsymbol{i}}^{-1} \hat{\boldsymbol{C}}.$$

The full demonstrations of Theorems 2.1 and 2.2 can be found, for example, in [BAG 10a].

REMARK 2.2.– The above-written estimators may be written in the form:

$$\hat{A}_j = \frac{U_j}{n}, \quad U_j = \sum_{i:X_i \in I_j} \delta_i, \quad \hat{\boldsymbol{C}}_j = \frac{1}{n} \sum_{i:X_i \in I_j} \delta_i \frac{\partial}{\partial \boldsymbol{\theta}} \ln \lambda(X_i; \hat{\boldsymbol{\theta}}).$$

$$E_j = \sum_{i:X_i > a_{j-1}} [\Lambda(a_j \wedge X_i; \hat{\boldsymbol{\theta}}) - \Lambda(a_{j-1}; \hat{\boldsymbol{\theta}})],$$

$$\hat{\boldsymbol{i}} = \frac{1}{n} \sum_{i=1}^{n} \delta_i \frac{\partial \ln \lambda(X_i; \hat{\boldsymbol{\theta}})}{\partial \boldsymbol{\theta}} \left(\frac{\partial \ln \lambda(X_i; \hat{\boldsymbol{\theta}})}{\partial \boldsymbol{\theta}}\right)^T.$$

Theorems 2.1 and 2.2 imply that a test for the hypothesis H_0 can be based on the statistic:

$$Y^2 = \boldsymbol{Z}^T \hat{\boldsymbol{V}}^- \boldsymbol{Z}, \qquad [2.30]$$

where $\hat{\boldsymbol{V}}^-$ is the general inverse of the matrix $\hat{\boldsymbol{V}}$. Using the expressions:

$$\hat{\boldsymbol{V}}^- = \hat{\boldsymbol{A}}^{-1} + \hat{\boldsymbol{A}}^{-1} \hat{\boldsymbol{C}}^T \hat{\boldsymbol{G}}^- \hat{\boldsymbol{C}} \hat{\boldsymbol{A}}^{-1}, \quad \hat{\boldsymbol{G}} = \hat{\boldsymbol{i}} - \hat{\boldsymbol{C}} \hat{\boldsymbol{A}}^{-1} \hat{\boldsymbol{C}}^T,$$

and the definition of $\boldsymbol{Z}$, the test statistic can be written in the form:

$$Y^2 = \sum_{j=1}^{k} \frac{(U_j - e_j)^2}{U_j} + Q, \qquad [2.31]$$

where

$$Q = \boldsymbol{W}^T \hat{\boldsymbol{G}}^{-} \boldsymbol{W}, \quad \boldsymbol{W} = \hat{\boldsymbol{C}} \hat{\boldsymbol{A}}^{-1} \boldsymbol{Z} = (W_1, ..., W_s)^T,$$

$$\hat{\boldsymbol{G}} = [\hat{g}_{ll'}]_{s \times s}, \quad \hat{g}_{ll'} = \hat{i}_{ll'} - \sum_{j=1}^{k} \hat{C}_{lj} \hat{C}_{l'j} \hat{A}_j^{-1},$$

$$W_l = \sum_{j=1}^{k} \hat{C}_{lj} \hat{A}_j^{-1} Z_j,$$

$$\hat{i}_{ll'} = \frac{1}{n} \sum_{i=1}^{n} \delta_i \frac{\partial \ln \lambda(X_i; \hat{\boldsymbol{\theta}})}{\partial \theta_l} \frac{\partial \ln \lambda(X_i; \hat{\boldsymbol{\theta}})}{\partial \theta_{l'}},$$

$$\hat{C}_{lj} = \frac{1}{n} \sum_{i: X_i \in I_j} \delta_i \frac{\partial}{\partial \theta_l} \ln \lambda(X_i; \hat{\boldsymbol{\theta}}), \quad j = 1, ..., k; \; l, l' = 1, ..., s. \tag{2.32}$$

By the well-known theorem on the distribution of quadratic forms ([GRE 96]), the limit distribution of the statistic Y_n^2 is chi-square with

$$r = rank \mathbf{V}^{-} = Tr(\mathbf{V}^{-}\mathbf{V})$$

degrees of freedom.

The hypothesis is rejected with the approximate significance level α if $Y_n^2 > \chi^2_\alpha(r)$.

If $\boldsymbol{G}$ is non-degenerate, then $r = k$ because using the equality $\boldsymbol{G} = \boldsymbol{i} - \boldsymbol{C}\boldsymbol{A}^{-1}\boldsymbol{C}^T$, we have:

$$\begin{aligned} \boldsymbol{V}^{-}\boldsymbol{V} &= (\boldsymbol{A}^{-1} + \boldsymbol{A}^{-1}\boldsymbol{C}^T\boldsymbol{G}^{-1}\boldsymbol{C}\boldsymbol{A}^{-1})(\boldsymbol{A} - \boldsymbol{C}^T\boldsymbol{i}^{-1}\boldsymbol{C}) \\ &= \mathbf{E} - \boldsymbol{A}^{-1}\boldsymbol{C}^T(\boldsymbol{i}^{-1} - \boldsymbol{G}^{-1}\boldsymbol{G}\boldsymbol{i}^{-1})\boldsymbol{C} = \mathbf{E}, \end{aligned}$$

so $Tr(\boldsymbol{V}^{-}\boldsymbol{V}) = k$.

REMARK 2.3.– (Hjort [HJO 90]) The matrix $\boldsymbol{G} = \boldsymbol{i} - \boldsymbol{C}\boldsymbol{A}^{-1}\boldsymbol{C}^T$ is degenerated if and only if the vector function $\psi(t; \boldsymbol{\theta}) = \frac{\partial}{\partial \boldsymbol{\theta}} \ln \lambda(t, \boldsymbol{\theta})$ satisfies the following condition: there exists a vector $\boldsymbol{x} = (x_1, ..., x_m) \neq (0, ..., 0)^T$ such that $\boldsymbol{x}^T \psi(t; \boldsymbol{\theta}_0)$ is constant on each interval I_j.

REMARK 2.4.– Replacing $\psi(t;\boldsymbol{\theta}_0)$ by $\psi(t;\hat{\boldsymbol{\theta}})$ and $y(u)\lambda(u;\boldsymbol{\theta}_0)du$ by $dN(u)$ in the expressions of $\boldsymbol{i}$, $\boldsymbol{C}$ and $\boldsymbol{A}$, we get $\hat{\boldsymbol{i}}$, $\hat{\boldsymbol{C}}$ and $\hat{\boldsymbol{A}}$, so we have that the matrix $\hat{\boldsymbol{G}} = \hat{\boldsymbol{i}} - \hat{\boldsymbol{C}}\hat{\boldsymbol{A}}^{-1}\hat{\boldsymbol{C}}^T$ is degenerated if and only if the vector function $\psi(t;\hat{\boldsymbol{\theta}})$ satisfies the following condition: there exists a vector $\boldsymbol{x} = (x_1, ..., x_m) \neq (0, ..., 0)^T$ such that for any fixed j, the value $\boldsymbol{x}^T\psi(X_i;\hat{\boldsymbol{\theta}})$ is the same and for all $X_i \in I_j$ such that $\delta_i = 1$.

REMARK 2.5.– The matrices $\boldsymbol{G}$ and $\hat{\boldsymbol{G}}$ are degenerated if there exists l such that $\frac{\partial}{\partial\theta_l}\ln\lambda(u,\boldsymbol{\theta})$ is constant on all intervals I_j. In such a case $g_{ll'} = \hat{g}_{ll'} = 0$ for all $l' = 1, ..., m$. For example, if the hazard function has the form:

$$\lambda(t;\theta,\nu) = \theta\, h(t,\nu), \quad \theta > 0, \nu \in \mathbf{R},$$

then $\psi(t;\theta,\nu) = \left(\theta^{-1}, (\ln h)'_\nu(t,\nu)\right)^T$, so taking $\boldsymbol{x} = (1,0)^T$, we have $\boldsymbol{x}^T\psi(t;\theta,\nu) = \theta^{-1}$. Then, $g_{11} = g_{12} = \hat{g}_{11} = \hat{g}_{12} = 0$.

2.4. The choice of grouping intervals

It is obvious that the power of the χ^2 tests for censored samples is also influenced by some other factors, associated with the form of censoring (the type and degree of censoring, as well as the distribution of censoring times). The usage of grouping methods, mentioned in Chapter 1, for censored samples may result in some difficulties, such as the fact that not all boundary points may belong to the observed interval $[0,\tau]$, or some intervals may not contain complete observations, what does not allow calculation of the NRR statistic. In such situations, it is necessary to combine intervals or change boundary points of intervals. In any case, the boundary points of grouping intervals should be chosen, taking into account the form of censoring. In this chapter, we consider three such grouping methods.

2.4.1. *Equifrequent grouping (EFG)*

It is recommended to take a_j as random data functions. In this grouping method, the boundary points are chosen in the interval $[0,\tau]$ from the condition of equal (or almost equal) numbers of complete observations, fallen into grouping intervals.

2.4.2. *Intervals with equal expected numbers of failures (EENFG)*

The idea is to divide the interval $[0, \tau]$ into k intervals with equal expected numbers of failures (which are not necessary integers). We think that it is better than to divide into intervals with equal estimated probabilities under the model because in many applications most of the right censoring times are concentrated at the right part of the data, so small number or no failures may be observed in the last intervals.

Define

$$E_k = \int_0^\tau \lambda(u; \hat{\boldsymbol{\theta}}) Y(u) du = \sum_{i=1}^n [\Lambda(X_i; \hat{\boldsymbol{\theta}}) - \Lambda(D_i; \hat{\boldsymbol{\theta}})],$$

$$E_j = \frac{j}{k} E_k, \quad j = 1, ..., k. \qquad [2.33]$$

So, we seek $\hat{a}_j$ verifying the equality

$$g(\hat{a}_j) = E_j, \quad g(a) = \int_0^a \lambda(u; \hat{\boldsymbol{\theta}}) Y(u) du. \qquad [2.34]$$

We shall prove that the limit distribution of the test statistic does not change.

We denote by $X_{(1)} \leq ... \leq X_{(n)}$ the ordered sample from $X_1, ..., X_n$. Note that the function

$$g(a) = \sum_{i=1}^n \Lambda(X_i \wedge a; \hat{\boldsymbol{\theta}})$$

$$= \sum_{i=1}^n \left[(n - i + 1) \Lambda(a; \hat{\boldsymbol{\theta}}) + \sum_{l=1}^{i-1} \Lambda(X_{(l)}; \hat{\boldsymbol{\theta}}) \right] \mathbf{1}_{[X_{(i-1)}, X_{(i)}]}(a)$$

is continuous and increasing on $[0, \tau]$; here $X_{(0)} = 0$, and we understand $\sum_{l=1}^0 c_l = 0$. Set

$$b_i = (n - i) \Lambda(X_{(i)}; \hat{\boldsymbol{\theta}}) + \sum_{l=1}^i \Lambda(X_{(l)}; \hat{\boldsymbol{\theta}}).$$

If i is the smallest natural number verifying $E_j \in [b_{i-1}, b_i]$, then the following equality can be written:

$$(n-i+1)\Lambda(a;\hat{\boldsymbol{\theta}}) + \sum_{l=1}^{i-1} \Lambda(X_{(l)};\hat{\boldsymbol{\theta}}) = E_j,$$

so if $E_j \in [b_{i-1}, b_i]$, $j = 1, ..., k-1$, then

$$\hat{a}_j = \Lambda^{-1}\left((E_j - \sum_{l=1}^{i-1} \Lambda(X_{(l)};\hat{\boldsymbol{\theta}}))/(n-i+1);\hat{\boldsymbol{\theta}} \right), \quad \hat{a}_k = X_{(n)}. \tag{2.35}$$

where Λ^{-1} is the inverse of the function Λ. We have $0 < \hat{a}_1 < \hat{a}_2 ... < \hat{a}_k = \tau$.

Under this choice of the intervals $e_j = E_k/k$ for any j.

Taking into account that:

$$\frac{E_j}{n} \xrightarrow{P} \frac{j}{k} \int_0^\tau \lambda(u;\boldsymbol{\theta}_0)y(u)du,$$

we have that $\hat{a}_j \xrightarrow{P} a_j$, where a_j is defined by:

$$\int_0^{a_j} \lambda(u;\boldsymbol{\theta}_0)y(u)du = \frac{j}{k} \int_0^\tau \lambda(t;\boldsymbol{\theta}_0)y(u)du.$$

By random change of time theorem (see [BIL 68]):

$$(H_n(\hat{a}_1), ..., H_n(\hat{a}_k))^T \xrightarrow{d} (V(a_1), ..., V(a_k))^T,$$

so replacing a_j with $\hat{a}_j$ in the expression of the statistic Y^2, the limit distribution of the statistic Y^2 is chi-square with r degrees of freedom, as in the case of fixed a_j.

2.4.3. *Optimal grouping (OptG)*

In the case of testing hypothesis H_0 against a given competing hypothesis H_1 by a randomly censored sample, the optimal boundary points of the grouping intervals can be obtained by solving the maximization problem, which is similar to [1.38]:

$$\max_{a_1 < a_2 < .... < a_{k-1}} \left(n \sum_{j=1}^{k} \frac{\left(p_j^1\left(\theta^1\right) - p_j^0\left(\theta^0\right) \right)^2}{p_j^0\left(\theta^0\right)} \right), \qquad [2.36]$$

where

$$p_j^0\left(\theta^0\right) = \int_{a_{j-1}}^{a_j} \left(1 - F^C(u)\right) f_0\left(u; \theta^0\right) du$$

and

$$p_j^1\left(\theta^1\right) = \int_{a_{j-1}}^{a_j} \left(1 - F^C(u)\right) f_1\left(u; \theta^1\right) du$$

are the probabilities of falling into j-th interval according to hypotheses H_0 and H_1, respectively. If the distribution of censoring times $F^C(t)$ is unknown (what is a typical situation in practice), then we can use the Kaplan-Meier estimator as $F^C(t)$, which is obtained by the inverted sample, in which δ_i is replaced by $1 - \delta_i$.

REMARK 2.6.– The maximized function in [2.36] is multiextremal, so global optimization methods are required to solve this problem.

REMARK 2.7.– Strictly speaking, the usage of EENFG and OptG does not guarantee that each grouping interval will contain complete observations, especially in the case of large degrees of censoring. In our simulation study, when such a situation occurs, we change boundary points, so that each interval would contain at least one complete observation.

Let us study the power of the considered χ^2 goodness-of-fit tests for a pair of close competing hypotheses through Monte Carlo simulations [CHI 14]. Three grouping methods, EFG, EENFG and OptG, are used, and the number of intervals k is taken equal to 3, 4, 5 and 6. Let us emphasize that all results, presented below, have been obtained in the case of testing composite hypotheses. The estimates of the power are calculated from the distributions of the test statistics $G(S|H_0)$ and $G(S|H_1)$, which are simulated based on censored samples of size $n = 200$. The number of simulations used is $N = 10^5$. The values of the power of the tests are calculated with the nominal significance level $\alpha = 0.1$. Let us consider first the null hypothesis H_0 corresponding to the Weibull distribution with the density function

$$f(t;\boldsymbol{\theta}) = \frac{\theta_2}{t}\left(\frac{t}{\theta_1}\right)^{\theta_2} \exp\left(-\left(\frac{t}{\theta_1}\right)^{\theta_2}\right)$$

and the parameters $\theta_1 = 2, \theta_2 = 2$. The competing hypothesis H_1 is the gamma distribution with the density function

$$f(t;\boldsymbol{\theta}) = \frac{1}{\theta_1 \cdot \Gamma(\theta_2)}\left(\frac{t}{\theta_1}\right)^{\theta_2 - 1} \exp\left(-\frac{t}{\theta_1}\right)$$

and the parameters $\theta_1 = 0.5577, \theta_2 = 3.1215$.

In the case of random censoring, we need to specify the distribution of censoring times. We have chosen two different families of distributions for censoring times: Type I beta-distribution family with the density function:

$$f^C(t;\theta) = B_I(\theta_1, \theta_2, \theta_3) = \frac{1}{\theta_3 \mathrm{B}(\theta_1, \theta_2)}\left(\frac{t}{\theta_3}\right)^{\theta_1 - 1}\left(1 - \frac{t}{\theta_3}\right)^{\theta_1 - 1},$$

where $\mathrm{B}(a, b)$ is the beta function and the Weibull distribution family. Censoring distributions are given in Table 2.1. The distribution parameters were adjusted so that the average degree of censoring under the considered null hypothesis is equal to some given value. In the case of Type I beta distributions, censored observations appear in the variational series of a censored sample approximately uniformly, as opposed to the Weibull distributions, for which censored observations appear generally at the end of the variational series.

In Table 2.2, the powers of the considered χ^2 tests are presented for various degrees of censoring in the case of randomly censored samples with censoring times from the Weibull distributions with parameter values, given in Table 1.2. In brackets, there is the number of intervals, for which the corresponding estimate of the power was obtained. In Table 2.3, the powers of the considered χ^2 tests are presented in the case of Type I beta distributions of censoring times.

Average degree of censoring	**Distribution of censoring times**	
10%	$B_I(1.81, 1, 7)$	W(3.44, 6.88)
20%	$B_I(1.19, 1, 7)$	W(2.87, 5.74)
30%	$B_I(1, 1.24, 7)$	W(2.48, 4.96)
40%	$B_I(1, 1.83, 7)$	W(2.16, 4.32)
50%	$B_I(1, 2.58, 7)$	W(1.87, 3.74)

Table 2.1. *Distributions of censoring times in the case of the lifetime distribution W(2, 2)*

		Average degree of censoring					
Test	**Grouping method**	0%	10%	20%	30%	40%	50%
NRR	EFG	0.51 (5)	0.38 (5)	0.32 (5)	0.28 (5)	0.25 (5)	0.22 (5)
	EENFG	0.51 (5)	0.41 (5)	0.34 (5)	0.30 (5)	0.27 (5)	0.23 (5)
	OptG	0.61 (6)	0.46 (4)	0.41 (4)	0.37 (4)	0.32 (3)	0.29 (3)
GPF	EFG	0.20 (6)	0.17 (6)	0.15 (6)	0.15 (6)	0.14 (6)	0.13 (6)
	EENFG	0.26 (6)	0.22 (6)	0.19 (6)	0.18 (6)	0.16 (6)	0.16 (6)
	OptG	0.56 (4)	0.51 (4)	0.44 (4)	0.37 (4)	0.35 (4)	0.29 (4)

Table 2.2. *Power of the tests for the Weibull-gamma pair in the case of randomly censored data with Weibull distribution of censoring times*

As can be seen from Tables 2.2 and 2.3, the power of the considered tests depends on the distribution of censoring times: in the case of Type I beta distribution, the power of the tests is higher than for the Weibull distribution of censoring times. The generalized Pearson-Fisher χ^2 test is considerably worse in terms of the power than the NRR test, when using EFG and EENFG. But in the case of optimal grouping, it is impossible to give preference to one test. The power of both tests in the case of optimal grouping is much higher than in the case of equifrequent grouping and grouping into intervals with equal expected numbers of failures.

Test	Grouping method	Average degree of censoring					
		0%	10%	20%	30%	40%	50%
NRR	EFG	0.51 (5)	0.43 (5)	0.40 (5)	0.36 (5)	0.33 (5)	0.28 (5)
	EENFG	0.51 (6)	0.46 (6)	0.42 (6)	0.38 (6)	0.34 (6)	0.30 (6)
	OptG	0.61 (6)	0.54 (6)	0.51 (4)	0.48 (4)	0.44 (4)	0.40 (4)
GPF	EFG	0.20 (6)	0.18 (6)	0.17 (6)	0.17 (6)	0.15 (6)	0.15 (6)
	EENFG	0.26 (6)	0.24 (6)	0.23 (6)	0.22 (6)	0.21 (6)	0.19 (6)
	OptG	0.56 (4)	0.53 (4)	0.49 (4)	0.45 (4)	0.41 (4)	0.37 (4)

Table 2.3. *Power of the tests for the Weibull-gamma pair in the case of randomly censored data with Type I beta distribution of censoring times*

2.5. Chi-squared tests for specific families of distributions

2.5.1. *Exponential distribution*

Let us consider the hypothesis:

$$H_0: \; F(t) = 1 - e^{-t/\theta}, \; t \geq 0$$

meaning that the distribution of the lifetimes is exponential; here $\theta > 0$ is an unknown parameter.

Under the null hypothesis:

$$S(t;\theta) = e^{-t/\theta}, \qquad \Lambda(t;\theta) = t/\theta, \qquad \lambda(t;\theta) = \frac{1}{\theta}, \qquad t \geq 0.$$

Note that we have the constant failure rate. In the class of all continuous distributions only the considered family of exponential distributions has this property. The constant failure rate function implies that there is no aging of items.

Set

$$S_n = \sum_{i=1}^{n} X_i, \quad \delta = \sum_{j=1}^{k} U_j, \quad U_j = \sum_{i:X_i \in I_j} \delta_i.$$

Then, we have:

$$\hat{\theta} = S_n/\delta, \quad \hat{C}_j = \frac{U_j \hat{\theta}}{n}, \quad \hat{i} = \frac{\delta \hat{\theta}^2}{n},$$

$$\hat{G} = \hat{g}_{11} = \frac{\delta \hat{\theta}^2}{n} - \sum_{j=1}^{k} \frac{U_j^2 \hat{\theta}^2}{n^2} \frac{n}{U_j} = 0.$$

Remarks 2.3–2.5 imply that $G = g_{11}$ and $\hat{G} = \hat{g}_{11}$ are degenerated, i.e. equal to zero. We cannot speak about the general inverse of $G = 0$ and $\hat{G} = 0$, so we find the general inverse $\hat{\boldsymbol{V}}^-$ of the matrix $\hat{\boldsymbol{V}}$ directly.

Under the exponential distribution, the elements of the matrix $\hat{\boldsymbol{V}}$ are:

$$\hat{v}_{jj} = \hat{A}_j - \hat{C}_j^2 \hat{i}^{-1} = \frac{U_j}{n} - \frac{U_j^2}{n\delta},$$

and for $j \neq j'$

$$v_{jj'} = -\hat{C}_j \hat{i}^{-1} \hat{C}_{j'} = -\frac{U_j U_{j'}}{n\delta}.$$

Set

$$\hat{\pi}_j = \frac{U_j}{\delta}, \quad \sum_{j=1}^{k} \hat{\pi}_j = 1, \quad \hat{\boldsymbol{\pi}} = (\hat{\pi}_1, ..., \hat{\pi}_k)^T.$$

Denote by $\boldsymbol{D}$ the diagonal matrix with the diagonal elements $\hat{\pi}$. The matrix $\hat{\boldsymbol{V}}$ and its generalized inverse $\hat{\boldsymbol{V}}^-$ have the form:

$$\hat{\boldsymbol{V}} = \frac{\delta}{n}(\hat{\boldsymbol{D}} - \hat{\boldsymbol{\pi}}\hat{\boldsymbol{\pi}}^T), \quad \hat{\boldsymbol{V}}^- = \frac{n}{\delta}(\hat{\boldsymbol{D}}^{-1} + \boldsymbol{1}\boldsymbol{1}^T),$$

because using the equalities:

$$\boldsymbol{1}^T \hat{\boldsymbol{D}} = \hat{\boldsymbol{\pi}}^T, \quad \boldsymbol{1}^T \hat{\boldsymbol{\pi}} = \hat{\boldsymbol{\pi}}^T \boldsymbol{1} = 1, \quad \hat{\boldsymbol{D}}\boldsymbol{1} = \hat{\boldsymbol{\pi}}, \quad \hat{\boldsymbol{\pi}}^T \hat{\boldsymbol{D}}^{-1} = \boldsymbol{1}^T,$$

we obtain the equality $\hat{\boldsymbol{V}}\hat{\boldsymbol{V}}^-\hat{\boldsymbol{V}} = \hat{\boldsymbol{V}}$.

The quadratic form $Y_n^2 = \boldsymbol{Z}^T \hat{\mathbf{V}}^- \boldsymbol{Z}$ has the form:

$$Y_n^2 = \frac{n}{\delta} \boldsymbol{Z}^T \hat{\boldsymbol{D}}^{-1} \boldsymbol{Z} + \frac{n}{\delta} (\boldsymbol{Z}^T \mathbf{1})^2 = \sum_{j=1}^{k} \frac{(U_j - e_j)^2}{U_j} + \frac{1}{\delta} \left[\sum_{j=1}^{k} (U_j - e_j) \right]^2 .$$

The limit distribution of the statistic Y_n^2 is chi-square with $Tr(\boldsymbol{V}^- \boldsymbol{V}) = k - 1$ degrees of freedom, because

$$\hat{A}_j \xrightarrow{P} A_j > 0, \quad \delta/n \xrightarrow{P} A = \sum_{j=1}^{k} A_j \in (0,1),$$

$$\hat{\pi}_j \xrightarrow{P} A_j / A = \pi_j, \quad \hat{\boldsymbol{D}} \xrightarrow{P} \boldsymbol{D},$$

so:

$$\boldsymbol{V}^- \boldsymbol{V} = \frac{1}{A} (\boldsymbol{D}^{-1} + \mathbf{1}\mathbf{1}^T)\, A\, (\boldsymbol{D} - \boldsymbol{\pi}\boldsymbol{\pi}^T) = \mathbf{E} - \mathbf{1}\boldsymbol{\pi}^T,$$

$$Tr(\mathbf{E} - \mathbf{1}\boldsymbol{\pi}^T) = k - \sum_{j=1}^{k} \pi_j = k - 1.$$

Note that:

$$\sum_{j=1}^{k} e_j = \frac{1}{\hat{\theta}} \int_0^T Y(u)du = \frac{1}{\hat{\theta}} \sum_{i=1}^{n} X_i = \frac{S_n}{\hat{\theta}} = \delta = \sum_{j=1}^{k} U_j .$$

$$Q = \frac{1}{\delta} \left[\sum_{j=1}^{k} (U_j - e_j) \right]^2 = 0.$$

Choice of $\hat{a}_j$. Set $S_0 = 0$, $S_i = (n - i) X_{(i)} + \sum_{l=1}^{i} X_{(l)}$, $i = 1, ..., n$. Formula [2.35] implies that the boundary points of the intervals I_j are chosen in the following way: if i is the smallest natural number verifying the inequalities

$$S_{i-1} \le \frac{j}{k} S_n \le S_i ,$$

then

$$\hat{a}_j = \frac{\frac{j}{k}S_n - \sum_{l=1}^{i-1} X_{(l)}}{n-i+1}, \quad j = 1, ..., k-1, \quad \hat{a}_k = X_{(n)}. \quad [2.37]$$

The number of expected failures in all intervals is equal:

$$e_j = \delta/k \quad \text{for any } j. \quad [2.38]$$

Chi-squared test for the exponential distribution: the null hypothesis is rejected with the approximate significance level α if

$$Y_n^2 = \sum_{j=1}^{k} \frac{(U_j - e_j)^2}{U_j} > \chi^2_\alpha(k-1).$$

EXAMPLE 2.1.– A total of 73 units were tested within $\tau = 109$ hours, but 44 units failed during the experiment. The failure times obtained are given in Table 2.4.

We want to test the hypothesis that the failure times have an exponential distribution.

i	X_i	δ_i	i	X_i	δ_i	i	X_i	δ_i	i	X_i	δ_i	i	X_i	δ_i	i	X_i	δ_i
1	3	1	14	27	1	27	54	1	40	94	1	53	109	0	66	109	0
2	3	1	15	30	1	28	60	1	41	95	1	54	109	0	67	109	0
3	5	1	16	32	1	29	63	1	42	99	1	55	109	0	68	109	0
4	5	1	17	33	1	30	64	1	43	101	1	56	109	0	69	109	0
5	6	1	18	35	1	31	64	1	44	103	1	57	109	0	70	109	0
6	8	1	19	37	1	32	72	1	45	109	0	58	109	0	71	109	0
7	10	1	20	41	1	33	74	1	46	109	0	59	109	0	72	109	0
8	10	1	21	41	1	34	77	1	47	109	0	60	109	0	73	109	0
9	11	1	22	46	1	35	81	1	48	109	0	61	109	0			
10	17	1	23	49	1	36	85	1	49	109	0	62	109	0			
11	20	1	24	49	1	37	87	1	50	109	0	63	109	0			
12	23	1	25	52	1	38	91	1	51	109	0	64	109	0			
13	24	1	26	52	1	39	91	1	52	109	0	65	109	0			

Table 2.4. *Type I censored sample*

The maximum likelihood estimate of parameter θ of exponential distribution $\hat{\theta} = 120.11$. For calculation of the test statistic Y_n^2, we have chosen the number of intervals $k = 5$. Boundary points a_j, E_j, numbers of observed failures U_j and expected numbers of failures e_j in the intervals I_j, $j = 1, \dots, 5$ are given in Table 2.5.

j	E_j	a_j	U_j	e_j	$\frac{(U_j-e_j)^2}{U_j}$
1	8.80	15.5625	9	8.80	0.004
2	17.60	32.9825	7	8.80	0.463
3	26.40	53.2340	10	8.80	0.144
4	35.20	77.7179	8	8.80	0.080
5	44.00	109.00	10	8.80	0.144

Table 2.5. *Intermediate calculation results*

Thus, $Y_n^2 = \sum\limits_{j=1}^{k} \frac{(U_j-e_j)^2}{U_j} = 0.835$. The limiting distribution is χ_4^2. The asymptotic p-value is equal to 0.93, so we have no reason to reject the hypothesis H_0 that times to failure have an exponential distribution.

2.5.2. *Weibull distribution*

Let us consider the hypothesis:

$$H_0 : F(t) = 1 - e^{-(t/\theta)^\nu}, \quad t \geq 0,$$

meaning that the distribution of the lifetimes is Weibull; here $\theta > 0, \ \nu > 0$ are unknown parameters.

Under the null hypothesis:

$$S(t;\theta,\nu) = \exp\left\{-\left(\frac{t}{\theta}\right)^\nu\right\}, \quad \Lambda(t;\theta,\nu) = \left(\frac{t}{\theta}\right)^\nu,$$

$$\lambda(t;\theta,\nu) = \frac{\nu}{\theta^\nu} t^{\nu-1}, \quad \ln\lambda(t;\theta,\nu) = (\nu-1)\ln t - \nu\ln\theta + \ln\nu.$$

Denote by $\hat{\theta}$ and $\hat{\nu}$ the ML estimators of the parameters θ and ν. These estimators maximize the log-likelihood function:

$$\ell(\theta,\nu) = \sum_{i=1}^{n} \{\delta_i[(\nu-1)\ln X_i - \nu\ln\theta + \ln\nu] - (X_i/\theta)^\nu\}.$$

Remarks 2.3–2.5 imply that the matrices $\boldsymbol{G}$ and $\hat{\boldsymbol{G}}$ are degenerated and have rank 1, $g_{11} = g_{12} = \hat{g}_{11} = \hat{g}_{12} = 0$. So, we need only $\hat{g}_{22}$ to find $\hat{G}^-$. We have:

$$\hat{i}_{22} = \frac{1}{n\hat{\nu}^2}\sum_{i=1}^{n}\delta_i[1+\ln Y_i]^2,$$

$$\hat{C}_{2j} = \frac{1}{n\hat{\nu}}\sum_{i:X_i\in I_j}\delta_i[1+\ln Y_i], \quad Y_i = \left(\frac{X_i}{\hat{\theta}}\right)^{\hat{\nu}},$$

$$\hat{g}_{22} = \hat{i}_{22} - \sum_{j=1}^{k}\hat{C}_{2j}^2\hat{A}_j^{-1}, \quad \hat{A}_j = U_j/n.$$

The matrix $\hat{\boldsymbol{G}}^-$ has the form:

$$\hat{\boldsymbol{G}}^- = \begin{pmatrix} 0 & 0 \\ 0 & \hat{g}_{22}^{-1} \end{pmatrix}.$$

Hence:

$$Q = \frac{W_2^2}{\hat{g}_{22}}, \quad W_2 = \sum_{j=1}^{k}\hat{C}_{2j}\hat{A}_j^{-1}Z_j, \quad Z_j = \frac{1}{\sqrt{n}}(U_j - e_j).$$

Choice of $\hat{a}_j$. Set

$$b_0 = 0, \quad b_i = (n-i)Y_{(i)} + \sum_{l=1}^{i}Y_{(l)}, \; i = 1, ..., n.$$

Formula [2.35] implies that the boundary points of the intervals I_j are chosen in the following way: if i is the smallest natural number verifying the inequalities

$$b_{i-1} \le \frac{j}{k}b_n \le b_i,$$

then

$$\hat{a}_j = \hat{\theta}\left(\frac{\frac{j}{k}b_n - \sum_{l=1}^{i-1} Y_{(l)}}{n-i+1}\right)^{1/\hat{\nu}}, \quad j = 1, ..., k-1, \quad \hat{a}_k = X_{(n)}. \tag{2.39}$$

In the case of such choice of intervals, we have:

$$e_j = \delta/k \quad \text{for all } j.$$

Chi-squared test for the Weibull distribution: the null hypothesis is rejected with the approximate significance level α if

$$Y_n^2 = \sum_{j=1}^{k} \frac{(U_j - e_j)^2}{U_j} + Q > \chi^2_\alpha(k-1).$$

EXAMPLE 2.2.– In Table 2.6, the lifetimes (in days) of $n = 168$ patients after surgical operation are given (see also [MAN 73], [GER 79]). We have the randomly censored sample with 66 censored observations.

In addition to the Weibull distribution, let us consider another distribution with a monotone hazard rate; the gamma distribution with density function:

$$f(t;\theta,\alpha) = \frac{1}{\theta \cdot \Gamma(\alpha)}\left(\frac{t}{\theta}\right)^{\alpha-1} \exp\left(-\frac{t}{\theta}\right), \quad t \geq 0, \theta, \alpha > 0.$$

Figure 2.1 shows the Kaplan-Meier estimate, obtained for the given sample of lifetimes, the survival functions corresponding to the Weibull distribution with ML estimates of parameters $\hat{\theta} = 1830.56$, $\hat{\nu} = 2.11$ and the gamma distribution with ML estimates of parameters $\hat{\theta} = 536.36, \hat{\alpha} = 3.14$. As you can see, these survival functions are very close to each other in this situation.

i	X_i	δ_i	i	X_i	δ_i	i	X_i	δ_i	i	X_i	δ_i	i	X_i	δ_i	i	X_i	δ_i
1	1102	1	29	377	1	57	1346	1	85	1977	1	113	2503	0	141	1873	1
2	952	1	30	1471	0	58	1340	1	86	620	1	114	868	1	142	727	0
3	2377	1	31	1793	1	59	953	1	87	1330	1	115	934	1	143	1289	1
4	1039	1	32	876	0	60	940	1	88	906	0	116	1105	1	144	797	1
5	1754	0	33	838	1	61	740	1	89	1023	0	117	1168	0	145	2027	0
6	1477	0	34	1179	1	62	1181	0	90	595	1	118	796	1	146	920	1
7	888	1	35	1120	1	63	55	1	91	717	0	119	886	0	147	1472	0
8	2531	1	36	1798	1	64	1400	0	92	947	1	120	1881	1	148	982	0
9	550	1	37	1855	0	65	1542	0	93	813	0	121	400	1	149	1322	1
10	973	0	38	1559	1	66	1130	0	94	473	1	122	974	1	150	1969	0
11	663	1	39	1681	1	67	609	1	95	965	1	123	1946	0	151	1564	1
12	1082	1	40	1701	1	68	474	1	96	1384	1	124	650	1	152	1417	1
13	2121	1	41	1374	0	69	1181	0	97	1793	0	125	808	1	153	974	1
14	1214	1	42	1039	1	70	1313	0	98	661	1	126	2190	1	154	1389	1
15	781	0	43	2421	0	71	164	1	99	1290	0	127	1108	1	155	2431	0
16	317	1	44	1537	1	72	1556	0	100	1500	0	128	1644	0	156	1577	0
17	1857	0	45	860	0	73	1269	1	101	1182	0	129	1230	0	157	1658	0
18	1646	0	46	470	0	74	655	1	102	1244	1	130	504	0	158	1755	0
19	1828	1	47	1644	0	75	1370	1	103	1122	0	131	2367	0	159	1199	1
20	1010	1	48	711	1	76	3302	0	104	2012	0	132	1444	0	160	1211	0
21	518	1	49	764	1	77	2285	0	105	1270	0	133	857	1	161	1725	0
22	1097	1	50	1453	0	78	1802	0	106	2186	1	134	1796	1	162	897	1
23	2123	1	51	2182	1	79	133	1	107	1362	1	135	1857	1	163	664	1
24	934	1	52	809	1	80	1435	1	108	2060	1	136	1636	0	164	1365	1
25	1693	1	53	1687	1	81	1588	0	109	1289	1	137	1677	0	165	1629	0
26	1182	1	54	1628	1	82	1560	0	110	699	0	138	1500	1	166	2322	1
27	2045	1	55	329	1	83	1373	1	111	2342	1	139	626	0	167	2075	0
28	1069	0	56	1746	1	84	855	1	112	1860	1	140	1165	1	168	2534	0

Table 2.6. *Lifetimes of patients after surgical operation*

So, let us test the hypothesis H_0: the lifetimes have the Weibull distribution, against H_1: the lifetimes have the gamma distribution. In this case, we can use the optimal grouping method. We have chosen the number of intervals $k = 4$. Then, solving the optimization problem [2.36], where $F^C(t)$ was taken as the Kaplan-Meier estimate from the inverted sample, in which δ_i is replaced by $1 - \delta_i$, we obtain the optimal grouping intervals.

Boundary points a_j, numbers of observed failures U_j and expected numbers of failures e_j in the intervals I_j, $j = 1, \ldots, 4$ are presented in Table 2.7.

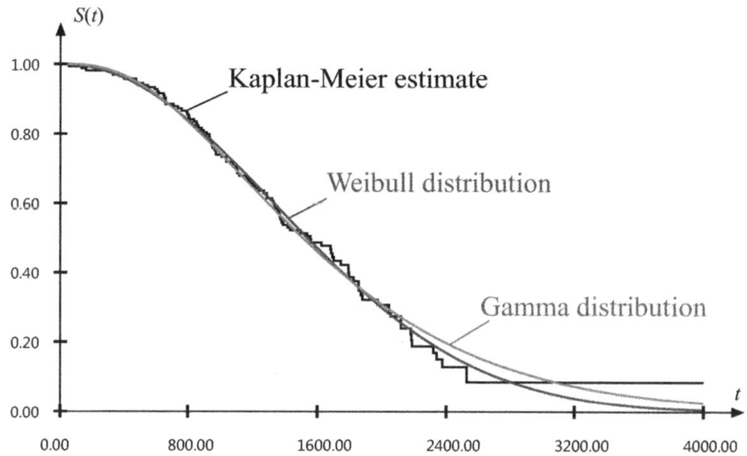

Figure 2.1. *Lifetime distribution. For a color version of this figure, see www.iste.co.uk/nikulin/chisquared.zip*

j	a_j	U_j	e_j	$\frac{(U_j-e_j)^2}{U_j}$
1	187.8645	3	1.36	0.897
2	353.0198	2	3.75	1.531
3	1294.3448	55	54.37	0.007
4	2531.00	42	42.51	0.006

Table 2.7. *Intermediate calculation results*

The following matrices were obtained and used for calculation of Q:

$$Z = (0.13, -0.14, 0.05, -0.04)^T,$$

$$\hat{C}_2 = (-0.0423, -0.0150, -0.0958, 0.1039).$$

Then, we obtain values $\hat{i}_{22} = 0.2331$, $\hat{g}_{22} = 0.0424$, and $W_2 = -0.1605$. Thus, $Q = 0.6073$, $\sum_{j=1}^{4} \frac{(U_j-e_j)^2}{U_j} = 2.4416$, and $Y_n^2 = 3.0489$. The limiting distribution is χ_3^2. The p-value is equal to 0.39, so we have no reason to reject the hypothesis that lifetimes have a Weibull distribution.

2.5.3. *Lognormal distribution*

Let us consider the hypothesis:

$$H_0 : F(t) = 1 - \Phi(\ln(t/\theta)^{\nu}), \quad t \geq 0,$$

meaning that the distribution of the lifetimes is lognormal; here Φ is the c.d.f. of standard normal distribution, $\theta > 0$ and $\nu > 0$ are unknown scalar parameters.

Under the null hypothesis:

$$S(t;\theta,\nu) = 1 - \Phi\left(\ln\left(\frac{t}{\theta}\right)^{\nu}\right), \quad \lambda(t;\theta,\nu) = \frac{\nu t^{\nu-1}}{\theta^{\nu}} \frac{\varphi\left(\ln\left(\frac{t}{\theta}\right)^{\nu}\right)}{1-\Phi\left(\ln\left(\frac{t}{\theta}\right)^{\nu}\right)}.$$

Denote by $\hat{\theta}$ and $\hat{\nu}$ the ML estimators of the parameters θ and ν. These estimators maximize the log-likelihood function:

$$\ell(\theta,\nu) = \sum_{i=1}^{n} \{\delta_i \left[\ln \nu - \ln X_i + V_i + \ln \frac{\varphi(V_i)}{1-\Phi(V_i)}\right] - \ln[1-\Phi(V_i)]\},$$

where $V_i = \ln(\frac{X_i}{\theta})^{\nu}$, φ is the p.d.f. of the standard normal distribution. The estimator $\hat{\boldsymbol{i}} = [\hat{i}_{ls}]_{2\times 2}$ has the form:

$$\hat{i}_{11} = \frac{\hat{\nu}^2}{n\hat{\theta}^2} \sum_{i=1}^{n} \delta_i g_1^2(Y_i), \quad \hat{i}_{12} = -\frac{1}{n\hat{\theta}} \sum_{i=1}^{n} \delta_i g_1(Y_i)[1+g_1(Y_i)\ln Y_i],$$

$$\hat{i}_{22} = \frac{1}{n\hat{\nu}^2} \sum_{i=1}^{n} \delta_i [1+g_1(Y_i)\ln Y_i]^2,$$

where

$$Y_i = \left(\frac{X_i}{\hat{\theta}}\right)^{\hat{\nu}}, \quad g_1(t) = \frac{\varphi(\ln t)}{1-\Phi(\ln t)} - \ln t.$$

The test statistic is:

$$Y_n^2 = \sum_{j=1}^{k} \frac{(U_j - e_j)^2}{U_j} + Q,$$

where

$$Q = \boldsymbol{W}^T \hat{\boldsymbol{G}}^{-1} \boldsymbol{W}, \quad \boldsymbol{W} = (W_1, W_2)^T,$$

$$\hat{\boldsymbol{G}} = [\hat{g}_{ll'}]_{2\times 2}, \quad \hat{g}_{ll'} = \hat{i}_{ll'} - \sum_{j=1}^{k} \hat{C}_{lj} \hat{C}_{l'j} \hat{A}_j^{-1},$$

$$W_l = \sum_{j=1}^{k} \hat{C}_{lj} \hat{A}_j^{-1} Z_j, \quad l, l' = 1, 2.$$

$$\hat{A}_j = U_j/n, \quad \hat{C}_{1j} = -\frac{\hat{\nu}}{n\hat{\theta}} \sum_{i:X_i \in I_j} \delta_i g_1(Y_i),$$

$$\hat{C}_{2j} = \frac{1}{n\hat{\nu}} \sum_{i:X_i \in I_j} \delta_i [1 + g_1(Y_i) \ln Y_i], \quad Z_j = \frac{1}{\sqrt{n}}(U_j - e_j).$$

Choice of $\hat{a}_j$. Set

$$b_i = (n-i)\Lambda_0(Y_i) + \sum_{l=1}^{i} \Lambda_0(Y_{(l)}),$$

where $\Lambda_0(t) = -\ln[1 - \Phi(\ln t)]$. If i is the smallest natural number verifying $E_j \in [b_{i-1}, b_i]$, then, for $j = 1, ..., k-1$,

$$\hat{a}_j = \hat{\theta} \left\{ \Lambda_0^{-1} \left((E_j - \sum_{l=1}^{i-1} \Lambda_0(Y_{(l)}))/(n-i+1) \right) \right\}^{1/\hat{\nu}}, \quad \hat{a}_k = X_{(n)},$$

where $\Lambda_0^{-1}(t) = \exp\{\Phi^{-1}(1 - e^{-t})\}$.

Under such choice of intervals we have $e_j = E_k/k$ for any j, where $E_k = \sum_{i=1}^{n} \Lambda_0(Y_i)$.

Chi-squared test for the lognormal distribution: the null hypothesis is rejected with the approximate significance level α if

$$Y_n^2 > \chi^2_\alpha(k).$$

EXAMPLE 2.3.– Let us consider the randomly censored sample, presented in Table 2.8. The sample size $n = 52$ and the number of censored observations is 15. It is necessary to test the hypothesis that the lifetimes have a lognormal distribution.

i	X_i	δ_i	i	X_i	δ_i	i	X_i	δ_i	i	X_i	δ_i
1	49	1	14	51	1	27	97	1	40	417	0
2	32	1	15	379	1	28	358	1	41	21	1
3	203	0	16	12	1	29	216	1	42	219	1
4	189	0	17	56	1	30	119	1	43	1225	1
5	76	1	18	124	1	31	538	0	44	14	1
6	300	0	19	47	1	32	348	1	45	943	0
7	162	1	20	68	1	33	469	1	46	23	1
8	478	0	21	165	1	34	84	1	47	530	0
9	37	1	22	32	1	35	267	0	48	222	1
10	437	1	23	156	1	36	331	0	49	149	1
11	82	1	24	57	1	37	29	1	50	171	1
12	1148	0	25	548	0	38	571	0	51	816	0
13	197	1	26	8	1	39	160	1	52	244	0

Table 2.8. *Randomly censored sample of lifetimes*

The ML estimates of parameters θ and ν of the lognormal distribution are $\hat{\theta} = 194.17$ and $\hat{\nu} = 1.58$. For calculation of the test statistic Y_n^2, we have chosen the number of intervals $k = 5$. The boundary points a_j, E_j, the observed numbers of failures U_j and the expected numbers of failures e_j in the intervals I_j, $j = 1, \ldots, 5$ are presented in Table 2.9.

j	E_j	a_j	U_j	e_j	$\frac{(U_j-e_j)^2}{U_j}$
1	7.4861	36.2689	8	7.49	0.033
2	14.9720	81.7371	8	7.49	0.033
3	22.4581	151.4903	6	7.49	0.368
4	29.9441	291.5963	9	7.49	0.255
5	37.4301	1225.00	6	7.49	0.368

Table 2.9. *Intermediate calculation results*

The following matrices were obtained and used for calculating Q:

$$Z = \begin{bmatrix} 0.0713 \\ 0.0713 \\ -0.2061 \\ 0.2099 \\ -0.2061 \end{bmatrix}, \quad \hat{A} = \begin{bmatrix} 0.1538 & 0 & 0 & 0 & 0 \\ 0 & 0.1538 & 0 & 0 & 0 \\ 0 & 0 & 0.1154 & 0 & 0 \\ 0 & 0 & 0 & 0.1730 & 0 \\ 0 & 0 & 0 & 0 & 0.1154 \end{bmatrix},$$

$$\hat{C} = \begin{bmatrix} -0.8E-03 & -0.5E-03 & -0.4E-03 & -0.5E-0.3 & -0.2E-03 \\ 0.14179 & -0.00316 & -0.04599 & -0.10635 & -0.07899 \end{bmatrix},$$

$$\hat{G} = \begin{bmatrix} 0.2E-06 & -0.2E-04 \\ -0.2E-04 & 0.08315 \end{bmatrix}, \quad W = \begin{bmatrix} -0.00019 \\ 0.15843 \\ 1.48125 \end{bmatrix}.$$

The values of the required estimates are:

$$\hat{\imath}_{11} = 0.9E-05, \quad \hat{\imath}_{12} = -0.2E-03, \quad \hat{\imath}_{22} = 0.08315.$$

Thus, $Q = W^T G^- W = 0.4244$, $\sum\limits_{j=1}^{5} \frac{(U_j - e_j)^2}{U_j} = 1.0568$ and $Y_n^2 = 1.4812$. The limiting distribution is χ_5^2. The p-value is equal to 0.91, so we have no reason to reject the hypothesis H_0.

2.5.4. *Loglogistic distribution*

Let us consider the hypothesis:

$$H_0: \; F(t) = 1 - \frac{1}{1 + (\frac{t}{\theta})^\nu}, \; t \geq 0,$$

meaning that the distribution of the lifetimes is loglogistic; here $\theta > 0, \nu > 0$ are unknown parameters.

Under the null hypothesis:

$$S(t;\theta,\nu) = \frac{1}{1+(\frac{t}{\theta})^{\nu}}, \quad \Lambda(t;\theta,\nu) = \ln\left[1+\left(\frac{t}{\theta}\right)^{\nu}\right],$$

$$\lambda(t;\theta,\nu) = \frac{\nu}{\theta^{\nu}}\frac{t^{\nu-1}}{1+(\frac{t}{\theta})^{\nu}}.$$

Denote by $\hat{\theta}$ and $\hat{\nu}$ the ML estimators of the parameters θ and ν. These estimators maximize the log-likelihood function:

$$\ell(\theta,\nu) = \sum_{i=1}^{n}\{\delta_i\left((\nu-1)\ln X_i - \nu\ln\theta\right.$$
$$\left.+\ln\nu - \ln[1+(X_i/\theta)^{\nu}]\right) - \ln[1+(X_i/\theta)^{\nu}]\}.$$

The estimator $\hat{\boldsymbol{i}} = [\hat{i}_{ls}]_{2\times 2}$ has the form:

$$\hat{i}_{11} = \frac{\hat{\nu}^2}{n\hat{\theta}^2}\sum_{i=1}^{n}\frac{\delta_i}{(1+Y_i)^2}, \quad \hat{i}_{12} = -\frac{1}{n\hat{\theta}}\sum_{i=1}^{n}\delta_i\frac{1+Y_i+\ln Y_i}{(1+Y_i)^2},$$

$$\hat{i}_{22} = \frac{1}{n\hat{\nu}^2}\sum_{i=1}^{n}\delta_i\left[1+\frac{\ln Y_i}{1+Y_i}\right]^2.$$

The test statistic is:

$$Y_n^2 = \sum_{j=1}^{k}\frac{(U_j-e_j)^2}{U_j} + Q,$$

where

$$Q = \boldsymbol{W}^T\hat{\boldsymbol{G}}^{-1}\boldsymbol{W}, \quad \boldsymbol{W} = (W_1,W_2)^T,$$

$$\hat{\boldsymbol{G}} = [\hat{g}_{ll'}]_{2\times 2}, \quad \hat{g}_{ll'} = \hat{i}_{ll'} - \sum_{j=1}^{k}\hat{C}_{lj}\hat{C}_{l'j}\hat{A}_j^{-1},$$

$$W_l = \sum_{j=1}^{k}\hat{C}_{lj}\hat{A}_j^{-1}Z_j, \quad l,l' = 1,2.$$

$$\hat{C}_{1j} = -\frac{\hat{\nu}}{n\hat{\theta}} \sum_{i:X_i \in I_j} \frac{\delta_i}{1+Y_i}, \quad \hat{C}_{2j} = \frac{1}{n\hat{\nu}} \sum_{i:X_i \in I_j} \delta_i \left(1 + \frac{\ln Y_i}{1+Y_i}\right).$$

Choice of $\hat{a}_j$. Set

$$Y_i = \left(\frac{X_i}{\hat{\theta}}\right)^{\hat{\nu}}, \quad b_i = (n-i)\ln(1+Y_{(i)}) + \sum_{l=1}^{i} \ln(1+Y_{(l)}), \quad E_j = \frac{j}{k} b_n.$$

Formula [2.35] implies that the limits of the intervals I_j are chosen in the following way: if i is the smallest natural number verifying the inequalities

$$b_{i-1} \leq E_j \leq b_i,$$

then

$$\hat{a}_j = \hat{\theta} \left[\exp\left\{ \left(E_j - \sum_{l=1}^{i-1} \ln(1+Y_{(l)}) \right) / (n-i+1) \right\} - 1 \right]^{1/\hat{\nu}},$$

$$j = 1, ..., k-1,$$

$$\hat{a}_k = X_{(n)}.$$

In the case of such choice of intervals, we have $e_j = b_n/k$ for any j.

Chi-squared test for the loglogistic distribution: the null hypothesis is rejected with the approximate significance level α if

$$Y_n^2 = \sum_{j=1}^{k} \frac{(U_j - e_j)^2}{U_j} + Q > \chi^2_\alpha(k).$$

EXAMPLE 2.4.– In Table 2.10, a randomly censored sample of lifetimes is given. The sample size $n = 117$ and the number of censored observations is 46. It is necessary to test the hypothesis that the lifetimes have a loglogistic distribution.

i	X_i	δ_i	i	X_i	δ_i	i	X_i	δ_i	i	X_i	δ_i	i	X_i	δ_i	i	X_i	δ_i
1	3166	1	21	4037	1	41	2153	1	61	5752	0	81	3477	0	101	3239	1
2	5356	0	22	3732	1	42	5313	0	62	5020	0	82	3344	1	102	2992	0
3	930	1	23	1919	1	43	1381	1	63	1564	1	83	3443	0	103	611	1
4	2209	1	24	4203	0	44	3487	1	64	3163	0	84	3576	0	104	3705	1
5	4768	0	25	2136	1	45	2299	1	65	417	1	85	1040	1	105	5452	0
6	4530	0	26	4944	0	46	4836	0	66	3305	0	86	1296	1	106	1221	1
7	3156	1	27	2211	1	47	4462	0	67	4142	0	87	3080	1	107	1580	1
8	3422	1	28	4437	0	48	830	1	68	2815	1	88	2309	0	108	4131	0
9	3984	1	29	2507	1	49	1824	1	69	2686	0	89	1509	1	109	2459	0
10	1477	1	30	4865	1	50	4783	0	70	1110	1	90	1909	1	110	3034	1
11	4357	0	31	4191	1	51	2969	1	71	3966	0	91	3118	1	111	2278	1
12	532	1	32	3562	1	52	577	1	72	3514	1	92	3699	0	112	3721	0
13	3507	1	33	3901	1	53	3759	0	73	2139	1	93	3001	0	113	2530	1
14	418	1	34	1866	1	54	5483	0	74	487	1	94	1590	1	114	1548	0
15	4045	1	35	3149	0	55	1243	1	75	3410	1	95	1436	1	115	3878	0
16	833	0	36	1697	0	56	4745	1	76	6221	0	96	1753	1	116	5098	0
17	4292	0	37	2188	1	57	4922	0	77	478	1	97	2471	1	117	2206	1
18	3423	0	38	4345	1	58	1263	1	78	1255	1	98	4230	0			
19	2224	1	39	445	1	59	2747	1	79	4047	0	99	3196	0			
20	473	1	40	1963	0	60	4115	0	80	3974	1	100	4366	1			

Table 2.10. *Randomly censored sample of lifetimes*

The ML estimates of parameters θ and ν of the loglogistic distribution are $\hat{\theta} = 3341.18$ and $\hat{\nu} = 1.92$. For calculation of the test statistic Y_n^2, we have set the number of intervals at $k = 5$. Boundary points a_j, E_j, number of observed failures U_j and expected numbers of failures e_j in the intervals I_j, $j = 1, \ldots, 5$ are presented in Table 2.11.

j	E_j	a_j	U_j	e_j	$\frac{(U_j-e_j)^2}{U_j}$
1	5.7243	593.6168	8	4.066	0.1179
2	11.4486	892.4776	2	4.376	1.3116
3	17.1729	1188.9049	3	5.478	0.5151
4	22.8972	1550.2514	9	7.548	0.3234
5	28.6215	6221.00	30	34.555	0.0001

Table 2.11. *Intermediate calculation results*

The following matrices were obtained and used to calculate Q:

$$Z = \begin{bmatrix} 0.3637 \\ -0.2197 \\ -0.2291 \\ 0.1342 \\ -0.4211 \end{bmatrix}, \quad \hat{A} = \begin{bmatrix} 0.0684 & 0 & 0 & 0 & 0 \\ 0 & 0.0171 & 0 & 0 & 0 \\ 0 & 0 & 0.0256 & 0 & 0 \\ 0 & 0 & 0 & 0.0769 & 0 \\ 0 & 0 & 0 & 0 & 0.2564 \end{bmatrix},$$

$$\hat{C} = \begin{bmatrix} -0.4E-04 & -0.9E-05 & -0.1E-04 & -0.4E-04 & -0.9E-04 \\ -0.09458 & -0.01622 & -0.09061 & -0.02001 & -0.06543 \end{bmatrix},$$

$$\hat{G} = \begin{bmatrix} 0.1E-07 & -0.2E-04 \\ -0.2E-04 & 0.06488 \end{bmatrix}, \quad W = \begin{bmatrix} 0.00013 \\ -0.31056 \end{bmatrix}.$$

The values of the required estimates are:

$$\hat{\imath}_{11} = 0.1E-06, \quad \hat{\imath}_{12} = 0.3E-0.4, \quad \hat{\imath}_{22} = 0.22408.$$

Thus, $Q = W^T G^- W = 1.7233$, $\sum\limits_{j=1}^{5} \frac{(U_j - e_j)^2}{U_j} = 7.7314$ and $Y_n^2 = 9.4547$. The limiting distribution is χ_5^2. The p-value is equal to 0.09, so the hypothesis H_0 is rejected with the approximate significance level $\alpha = 0.1$.

2.5.5. *Gompertz distribution*

Let us consider the hypothesis:

$$H_0 : F(t) = 1 - \exp\left\{-\frac{\theta}{\nu}(e^{\nu t} - 1)\right\}, \; t \geq 0,$$

meaning that the distribution of the lifetimes is Gompertz; here $\theta > 0$, $\nu > 0$ are unknown parameters.

Under the null hypothesis:

$$\lambda(t; \theta, \nu) = \theta e^{\nu t}, \quad \Lambda(t; \theta, \nu) = \frac{\theta}{\nu}(e^{\nu t} - 1).$$

The log-likelihood function is:

$$\ell(\theta,\nu)=\sum_{i=1}^{n}\{\delta_i[\ln\theta+\nu X_i]-\frac{\theta}{\nu}(e^{\nu X_i}-1)\}.$$

Denote by $\hat{\theta}$ and $\hat{\nu}$ the ML estimators of θ and ν.

Remarks 2.3–2.5 imply that the matrices $\boldsymbol{G}$ and $\hat{\boldsymbol{G}}$ are degenerated and have rank 1, $g_{11}=g_{12}=\hat{g}_{11}=\hat{g}_{12}=0$. So, we need only $\hat{g}_{22}$ to find $\hat{G}^{-}$. We have:

$$\hat{\imath}_{22}=\frac{1}{n}\sum_{i=1}^{n}\delta_i X_i^2,\quad \hat{C}_{2j}=\frac{1}{n}\sum_{i:X_i\in I_j}\delta_i X_i,$$

$$\hat{g}_{22}=\hat{\imath}_{22}-\sum_{j=1}^{k}\hat{C}_{2j}^2\hat{A}_j^{-1},\quad \hat{A}_j=U_j/n.$$

Hence:

$$Q=\frac{W_2^2}{\hat{g}_{22}},\quad W_2=\sum_{j=1}^{k}\hat{C}_{2j}\hat{A}_j^{-1}Z_j,\quad Z_j=\frac{1}{\sqrt{n}}(U_j-e_j).$$

Choice of $\hat{a}_j$. Set

$$b_0=0,\quad b_i=(n-i)\frac{e^{\hat{\nu}X_{(i)}}-1}{\hat{\nu}}+\sum_{l=1}^{i}\frac{e^{\hat{\nu}X_{(l)}}-1}{\hat{\nu}},$$

$$i=1,...,n-1,\quad \hat{a}_n=X_{(n)}.$$

Formula [2.35] implies that the limits of the intervals I_j are chosen in the following way: if i is the smallest natural number verifying the inequalities

$$b_{i-1}\le\frac{j}{k}b_n\le b_i,$$

then

$$a_j = \frac{1}{\hat{\nu}} \ln \left\{ 1 + \hat{\nu} (\frac{j}{k} b_n - \sum_{l=1}^{i-1} \frac{e^{\hat{\nu} X_{(l)}} - 1}{\hat{\nu}}) / (n - i + 1) \right\},$$

$$\hat{j} = 1, ..., k - 1, \quad \hat{a}_k = X_{(n)}.$$

In the case of such a choice of intervals, we have $e_j = \hat{\theta} b_n / k$.

Chi-squared test for the Gompertz distribution: the null hypothesis is rejected with the approximate significance level α if

$$Y_n^2 = \sum_{j=1}^{k} \frac{(U_j - e_j)^2}{U_j} + Q > \chi^2_\alpha (k - 1).$$

EXAMPLE 2.5.– $n = 120$ electronic devices were observed during the time $\tau = 5.54$ (years). The number of failures is equal to 113. The failure times are given in Table 2.12 (see also [BAG 10b]). We want to test the hypothesis that the failure times have a Gompertz distribution.

The ML estimates of the parameters θ and ν are $\hat{\theta} = 0.00625$ and $\hat{\nu} = 1.10058$. For calculation of the test statistic Y_n^2, we have chosen the number of intervals $k = 10$. Boundary points a_j, numbers of observed failures U_j and expected numbers of failures e_j in the intervals I_j, $j = 1, \ldots, 10$ are presented in Table 2.13. The values of required estimators are:

$$\hat{g}_{22} = 0.30322, \quad \hat{i}_{22} = 17.0672, \quad W_2 = -0.32875.$$

The value of the test statistic is $Y_n^2 = X_n^2 + Q = 10.3323 + 0.3564 = 10.6887$. The limiting distribution is χ^2_9. The p-value is equal to 0.2977, so we have no reason to reject the hypothesis that failure times have a Gompertz distribution.

i	X_i	δ_i	i	X_i	δ_i	i	X_i	δ_i	i	X_i	δ_i	i	X_i	δ_i	i	X_i	δ_i
1	1.744	1	21	3.3507	1	41	3.7157	1	61	4.2885	1	81	4.8406	1	101	5.1816	1
2	1.9172	1	22	3.3514	1	42	3.7288	1	62	4.2919	1	82	4.8532	1	102	5.2441	1
3	2.1461	1	23	3.3625	1	43	3.7502	1	63	4.297	1	83	4.8619	1	103	5.2546	1
4	2.3079	1	24	3.3802	1	44	3.7823	1	64	4.3666	1	84	4.8635	1	104	5.3353	1
5	2.3753	1	25	3.3855	1	45	3.8848	1	65	4.3918	1	85	4.8679	1	105	5.4291	1
6	2.3858	1	26	3.4012	1	46	3.8902	1	66	4.4365	1	86	4.8858	1	106	5.436	1
7	2.4147	1	27	3.4382	1	47	3.9113	1	67	4.4919	1	87	4.8928	1	107	5.4633	1
8	2.5404	1	28	3.4438	1	48	3.9468	1	68	4.4932	1	88	4.9466	1	108	5.4842	1
9	2.6205	1	29	3.4684	1	49	3.9551	1	69	4.5388	1	89	4.9846	1	109	5.486	1
10	2.6471	1	30	3.5019	1	50	3.9728	1	70	4.5826	1	90	5.0008	1	110	5.4903	1
11	2.837	1	31	3.511	1	51	3.9787	1	71	4.5992	1	91	5.0144	1	111	5.5199	1
12	2.8373	1	32	3.5297	1	52	3.9903	1	72	4.6001	1	92	5.0517	1	112	5.5232	1
13	2.8766	1	33	3.5363	1	53	4.0078	1	73	4.6324	1	93	5.0898	1	113	5.5335	1
14	2.9888	1	34	3.5587	1	54	4.0646	1	74	4.64	1	94	5.0929	1	114	5.54	0
15	3.072	1	35	3.5846	1	55	4.1301	1	75	4.7164	1	95	5.0951	1	115	5.54	0
16	3.1586	1	36	3.5992	1	56	4.1427	1	76	4.73	1	96	5.1023	1	116	5.54	0
17	3.173	1	37	3.654	1	57	4.23	1	77	4.7881	1	97	5.1219	1	117	5.54	0
18	3.2132	1	38	3.6574	1	58	4.2312	1	78	4.7969	1	98	5.1223	1	118	5.54	0
19	3.2323	1	39	3.6674	1	59	4.2525	1	79	4.8009	1	99	5.171	1	119	5.54	0
20	3.3492	1	40	3.7062	1	60	4.2581	1	80	4.8351	1	100	5.1766	1	120	5.54	0

Table 2.12. *Type I censored sample of failure times*

j	a_j	U_j	e_j	$\frac{(U_j-e_j)^2}{U_j}$
1	2.6254	9	11.3	0.5878
2	3.2740	10	11.3	0.1690
3	3.7022	20	11.3	3.7845
4	4.0461	14	11.3	0.5207
5	4.3316	10	11.3	0.1690
6	4.5803	6	11.3	4.6817
7	4.8060	10	11.3	0.1690
8	5.0430	12	11.3	0.0408
9	5.3304	12	11.3	0.0408
10	5.8982	10	11.3	0.1690

Table 2.13. *Intermediate calculation results*

3

Comparison of the Chi-squared Goodness-of-fit Test with Other Tests

In this chapter, we discuss some advantages and disadvantages of chi-squared goodness-of-fit tests and compare them with tests based on the difference between non-parametric and parametric estimators, such as the Kolmogorov, Cramer-von Mises-Smirnov and Anderson-Darling tests. We carry out the investigation by the Monte Carlo method.

3.1. Tests based on the difference between non-parametric and parametric estimators

Let us consider a *complete sample* $\mathbf{X} = (X_1, X_2, ..., X_n)^T$ of independent identically distributed random variables. One approach for testing goodness-of-fit hypothesis

$$H_0 : F(t) \in \{F(t;\boldsymbol{\theta}), \boldsymbol{\theta} \in \Theta \subset \mathbf{R}^s\}$$

by complete samples is the application of non-parametric tests, such as Kolmogorov, Cramer-von Mises-Smirnov, Anderson-Darling and Zhang tests. The Kolmogorov test statistic is given by:

$$D_n = \sup_{-\infty<t<\infty} |F_n(t) - F(t;\boldsymbol{\theta})|, \qquad [3.1]$$

where $F_n(t)$ is the empirical distribution function. In practice, the statistic is usually used with the Bolshev correction [BOL 83] of the form:

$$S_K = \frac{6n \cdot D_n + 1}{6\sqrt{n}}, \qquad [3.2]$$

where $D_n = \max(D_n^+, D_n^-)$,

$$D_n^+ = \max_{1 \le i \le n} \left\{ \frac{i}{n} - F(X_{(i)}; \boldsymbol{\theta}) \right\}, D_n^- = \max_{1 \le i \le n} \left\{ F(X_{(i)}; \boldsymbol{\theta}) - \frac{i-1}{n} \right\},$$

n is the sample size and $X_{(1)}, X_{(2)}, \ldots, X_{(n)}$ are the order statistics (sample values in increasing order). The Cramer-von Mises-Smirnov test statistic is defined as follows:

$$S_{\omega^2} = n \int_{-\infty}^{\infty} (F_n(t) - F(t; \boldsymbol{\theta}))^2 \, dF(t; \boldsymbol{\theta}). \qquad [3.3]$$

After taking the integral, we have:

$$S_{\omega^2} = \frac{1}{12n} + \sum_{i=1}^{n} \left\{ F(X_{(i)}; \boldsymbol{\theta}) - \frac{2i-1}{2n} \right\}^2. \qquad [3.4]$$

The Anderson-Darling test statistic is defined as:

$$S_{\Omega^2} = n \int_{-\infty}^{\infty} (F_n(t) - F(t; \boldsymbol{\theta}))^2 \frac{dF(t; \boldsymbol{\theta})}{F(t; \boldsymbol{\theta})(1 - F(t; \boldsymbol{\theta}))}. \qquad [3.5]$$

After taking the integral, it can be presented as:

$$S_{\Omega^2} = -n - 2\sum_{i=1}^{n} \left\{ \frac{2i-1}{2n} \ln F(X_{(i)}; \boldsymbol{\theta}) + \left(1 - \frac{2i-1}{2n}\right) \ln(1 - F(X_{(i)}; \boldsymbol{\theta})) \right\}. \qquad [3.6]$$

In [ZHA 02] and [ZHA 05], Zhang proposed goodness-of-fit tests, based on the following statistics:

$$Z_A = -\sum_{i=1}^{n}\left[\frac{\ln\{F(X_{(i)};\boldsymbol{\theta})\}}{n-i+\frac{1}{2}} + \frac{\ln\{1-F(X_{(i)};\boldsymbol{\theta})\}}{i-\frac{1}{2}}\right], \quad [3.7]$$

$$Z_C = \sum_{i=1}^{n}\left[\ln\left\{\frac{[F(X_{(i)};\boldsymbol{\theta})]^{-1}-1}{(n-\frac{1}{2})/(i-\frac{3}{4})-1}\right\}\right]^2, \quad [3.8]$$

$$Z_K = \max_{1\le i\le n}\left(\left(i-\frac{1}{2}\right)\ln\left\{\frac{i-\frac{1}{2}}{nF(X_{(i)};\boldsymbol{\theta})}\right\} + \left(n-i+\frac{1}{2}\right)\ln\left[\frac{n-i+\frac{1}{2}}{n}(1-F(X_{(i)};\boldsymbol{\theta}))\right]\right). \quad [3.9]$$

Let us denote by $G(s|H_0)$ the distribution of a test statistic under hypothesis H_0. In the case of testing simple hypotheses, the distributions $G(s|H_0)$ of the statistics considered do not depend on the tested distribution. In the limit with $n \to \infty$, the statistic S_K belongs to the Kolmogorov distribution, and S_ω and S_Ω belong to the $a1$ and $a2$ distributions, respectively [BOL 83]. The distributions of Zhang's statistics depend on the sample size n, which strongly complicates the application of these tests in practice. For composite hypotheses, the test statistic distributions $G(s|H_0)$ are affected by a number of factors: the form of the tested lifetime distribution $F(t;\boldsymbol{\theta})$, the type and the number of estimated parameters, the method of parameter estimation and other factors. Approximations of the limiting statistic distributions for testing various composite hypotheses have been proposed in [LEM 09a] and [LEM 09b]. The distributions of Zhang's statistics are dependent on the sample size.

Now, let us consider the modification of the classical Kolmogorov, Cramer-von Mises-Smirnov and Anderson-Darling tests for *censored samples* of the form:

$$\mathbf{X} = (X_1, \delta_1), (X_2, \delta_2), ..., (X_n, \delta_n),$$

where $X_i = \min(T_i, C_i)$ is the observation value, T_i is the lifetime and C_i is a censoring time. The censoring indicator $\delta_i = I\{T_i \le C_i\}$.

The modified Kolmogorov, Cramer-von Mises-Smirnov and Anderson-Darling tests for type I and II censored samples are given in papers [BAR 73, D'AG 86, REI 04, BOR 84, COH 91, CAS 11, AGU 02, KAL 85, TIT 08]. In [LEM 10a], the authors proposed approximations of the limiting distributions of the test statistics for type I and II censored samples and carried out the power study.

In the case of randomly censored data, these tests can be modified by using the Kaplan-Meier estimate instead of the empirical distribution function $F_n(t)$ in the formulas of statistics [3.1], [3.3] and [3.5] (see, for example, [HOL 92], [KOZ 76], [NAI 81] and [REI 04]). In [CHI 11c], [LEM 10a] and [CHI 11a], the distributions of the modified Kolmogorov, Cramer-von Mises-Smirnov and Anderson-Darling statistics were investigated by means of the Monte Carlo simulation method. In this case, these statistics are calculated on the interval $[0, \tau]$, where $\tau = \max\limits_{1 \le i \le n} \{\delta_i X_i\}$ is the last observed failure time.

In [CHI 11c], [LEM 10a] and [CHI 11a], we investigated the distributions of the modified statistics for type I, type II and randomly censored samples. The distributions of the modified test statistics considerably depend on the type of censoring and censoring degree. In the case of randomly censored samples, distributions of test statistics considered also depend on the distribution of censoring times, which is usually unknown in practice. By this reason, we propose to use the Kaplan-Meier estimate of the distribution of censoring times for simulation of a randomly censored sample.

So, in order to simulate the same censoring scheme as in the original sample, we can use the following algorithm:

1) Generate a complete sample of the size n according to the distribution $F(t; \boldsymbol{\theta})$: $T_i = F^{-1}(\zeta_i; \boldsymbol{\theta})$, where ζ_i is uniformly distributed at the interval $U(0, 1)$, $i = 1, 2, ..., n$.

2) Inverse the original censored sample by changing the values δ_i on $1 - \delta_i$.

3) Calculate the Kaplan-Meier estimate of the distribution of censoring time $\hat{F}^C(t)$ by the inversed sample obtained from the previous step.

4) Generate ξ_i uniformly distributed at the interval $U(0,1)$ and calculate the values C_i, $i = 1, 2, ..., n$:

a) if $\xi_i < \hat{F}^C(c_1)$, then $C_i = \frac{\xi_i \cdot c_1}{\hat{F}^C(c_1)}$;

b) if $\xi_i \in \left(\hat{F}^C(c_j), \hat{F}^C(c_{j+1})\right]$, then

$C_i = c_j + \frac{(\xi_i - \hat{F}^C(c_j)) \cdot (c_{j+1} - c_j)}{\left(\hat{F}^C(c_{j+1}) - \hat{F}^C(c_j)\right)}$, $j = 1, 2, ..., r$;

c) if $\xi_i > \hat{F}^C(c_r)$, then $C_i = c_r + c_r\left(\xi_i - \hat{F}^C(c_r)\right)$, where $c_1, ..., c_r$ are the increase-ordered different censoring times in the original sample and r is the number of different censoring times in the original sample.

5) $X_i = \min(T_i, C_i)$, $\delta_i = 1\ \{T_i \leq C_i\}$, $i = 1, 2, ..., n$.

So, for testing goodness-of-fit with the modified tests and calculating the p-value, it is necessary to simulate the statistic distribution under the null hypothesis. However, the process of simulation requires time and computational costs and it is not possible without special software. This is a significant disadvantage in the modified Kolmogorov, Cramer-von Mises-Smirnov and Anderson-Darling tests for censored samples. Moreover, in the case of random censoring, the distribution of modified test statistic depends on the censoring degree, which in turn is defined by the combination of distributions $F(t;\boldsymbol{\theta})$ and $F^C(t)$. If the distribution $F(t;\boldsymbol{\theta})$, corresponding to the tested hypothesis H_0, is close enough to the unknown true distribution $F(t)$, then the average censoring degree of simulated samples will be close to the censoring degree of the original sample; otherwise, the application of the modified tests will be incorrect.

In [CHI 11b], the authors proposed an approach, which makes it possible to apply the classical Kolmogorov, Cramer-von Mises-Smirnov and Anderson-Darling tests with statistics [3.2], [3.4] and

[3.6] for the censored sample. The approach is based on transforming the original censored sample into a complete sample by using randomization. For each censored observation $(X_i,\ \delta_i = 0)$, a value $\hat{T}_i$ is generated by $\hat{T}_i = F^{-1}(\xi_i; \boldsymbol{\theta})$, where $\xi_i \sim U\left[F\left(C_i; \boldsymbol{\theta}\right), 1\right)$. The corresponding values of censored observations in the original sample are replaced by generated values. As a result, we obtain a pseudo-complete sample $\hat{X}_1, \hat{X}_2, ..., \hat{X}_n$, in which

$$\hat{X}_i = \begin{cases} X_i, & \text{if } \delta_i = 1, \\ \hat{T}_i, & \text{if } \delta_i = 0. \end{cases}$$

However, this approach is really good only in the case of simple hypothesis, when the parameter $\boldsymbol{\theta}$ is known. In [CHI 11b], it has been shown that under the simple hypothesis H_0, the distribution of the completed sample is $F(t; \boldsymbol{\theta})$. So, the algorithm of testing simple hypotheses can be written as follows:

1) Specify the significance level α;

2) Transform the original censored sample into a pseudo-complete sample $\hat{X}_1, \hat{X}_2, ..., \hat{X}_n$;

3) Calculate the test statistic S^* ([3.2], [3.4] or [3.6]) from this pseudo-complete sample and the specified distribution $F(t; \boldsymbol{\theta})$;

4) Compute the p-value as $p = 1 - G(S^*|H_0)$;

5) The hypothesis H_0 is rejected if the obtained p-value is less than the pre-specified α.

Through Monte Carlo simulations, it was shown (see [CHI 11b]) that the distributions of the test statistics based on the pseudo-complete samples, when testing the simple hypothesis, do not depend on the distribution $F^C(t)$ and coincide with the corresponding limiting distributions (for example, the Kolmogorov distribution for Kolmogorov's test statistic).

3.2. Comparison of goodness-of-fit tests for complete data

In [LEM 09c] and [LEM 10b], the Kolmogorov, Cramer-von Mises-Smirnov and Anderson-Darling tests as well as the chi-squared

Pearson and Nikulin-Rao-Robson tests were compared by power in the case of testing simple and composite hypotheses from complete data. In [LEM 15b], the detailed analysis of advantages and disadvantages of various goodness-of-fit tests as well as the special normality tests is given. A comparative analysis of the tests considered for the inverse Gaussian distribution was carried out in [LEM 11a].

Let us compare the tests considered in terms of power when testing H_0, corresponding to the normal distribution, against the competing hypotheses, which were considered in section 1.7. The density functions are shown in Figure 1.2. The estimates of the test power for the competing hypothesis H_1, corresponding to the generalized normal distribution, are given in Table 3.1. In Table 3.2, the estimates of the test power are provided for the competing hypothesis H_2, corresponding to the Laplace distribution, and in Table 3.3, they are given for hypothesis H_3, corresponding to the logistic distribution. The powers of the chi-squared tests were calculated using the asymptotically D-optimal grouping. Unknown parameters of the normal distribution were estimated by the maximum likelihood method from the initial complete sample. The number of simulations is $N = 10^5$. The values of the power of the tests are calculated for the significance level $\alpha = 0.1$.

	$n = 20$	$n = 30$	$n = 40$	$n = 50$	$n = 100$	$n = 150$	$n = 200$	$n = 300$
X_n^2	0.146 (4)	0.216 (5)	0.267 (5)	0.311 (5)	0.479 (5)	0.634 (7)	0.695 (9)	0.858 (11)
Y_n^2	0.107 (4)	0.197 (5)	0.221 (5)	0.24 (5)	0.395 (5)	0.539 (7)	0.616 (9)	0.81(11)
Z_A	0.09	0.128	0.18	0.239	0.548	0.775	0.901	0.985
Z_C	0.13	0.179	0.238	0.3	0.599	0.806	0.917	0.988
Z_K	0.13	0.144	0.164	0.186	0.334	0.5	0.651	0.856
S_{Ω^2}	0.148	0.192	0.238	0.287	0.526	0.719	0.848	0.964
S_{ω^2}	0.149	0.187	0.227	0.269	0.477	0.656	0.79	0.934
S_K	0.131	0.155	0.181	0.208	0.35	0.49	0.618	0.808

Table 3.1. *Test power for the normal-generalized normal pair when testing the composite hypothesis*

As can be seen from Tables 3.1–3.3, in most situations, the Zhang tests with statistics Z_A and Z_C turned out to be more powerful than the other tests considered. However, for small sample sizes in the case of competing hypothesis H_1, Zhang's test with statistic Z_A yielded in power to other tests. The Anderson-Darling test is one of the most preferable tests by power.

	$n = 20$	$n = 30$	$n = 40$	$n = 50$	$n = 100$	$n = 150$	$n = 200$	$n = 300$
X_n^2	0.177 (4)	0.261 (5)	0.314 (5)	0.381 (5)	0.708 (9)	0.86 (9)	0.946 (11)	0.993 (13)
Y_n^2	0.173 (4)	0.262 (5)	0.329 (5)	0.409 (5)	0.755 (9)	0.889 (9)	0.961 (11)	0.995 (11)
Z_A	0.368	0.449	0.518	0.578	0.793	0.908	0.963	0.995
Z_C	0.331	0.413	0.485	0.548	0.773	0.896	0.956	0.992
Z_K	0.326	0.424	0.504	0.569	0.788	0.901	0.955	0.992
S_{Ω^2}	0.347	0.455	0.549	0.63	0.879	0.966	0.991	1
S_{ω^2}	0.336	0.444	0.537	0.621	0.875	0.965	0.991	1
S_K	0.302	0.388	0.468	0.54	0.797	0.92	0.972	0.997

Table 3.2. *Test power for the normal-Laplace pair when testing the composite hypothesis*

	$n = 20$	$n = 30$	$n = 40$	$n = 50$	$n = 100$	$n = 150$	$n = 200$	$n = 300$
X_n^2	0.125 (4)	0.125 (5)	0.134 (6)	0.143 (5)	0.220 (9)	0.284 (10)	0.355 (11)	0.486 (13)
Y_n^2	0.12 (4)	0.137 (5)	0.156 (6)	0.172 (6)	0.279 (9)	0.358 (10)	0.436 (11)	0.572 (13)
Z_A	0.198	0.228	0.252	0.272	0.358	0.435	0.507	0.633
Z_C	0.186	0.217	0.245	0.27	0.374	0.46	0.537	0.662
Z_K	0.162	0.195	0.225	0.249	0.349	0.43	0.499	0.614
S_{Ω^2}	0.163	0.187	0.21	0.23	0.327	0.415	0.498	0.639
S_{ω^2}	0.152	0.172	0.191	0.209	0.294	0.373	0.449	0.585
S_K	0.143	0.156	0.169	0.181	0.24	0.297	0.354	0.462

Table 3.3. *Test power for the normal-logistic pair when testing the composite hypothesis*

Despite the fact that the chi-squared tests showed lower power in comparison with the Zhang and Anderson-Darling tests, when testing the composite hypothesis, we should not completely reject the application of these tests. First, the chi-squared Nikulin-Rao-Robson test is much simpler to use, as the limiting distribution of the statistic Y_n^2 is the simple χ^2_{k-1} distribution, when using ML estimates of unknown parameters from the complete sample (see Chapter 1). Second, it is possible to maximize the power of chi-squared tests for a given pair of competing hypotheses using the optimal grouping.

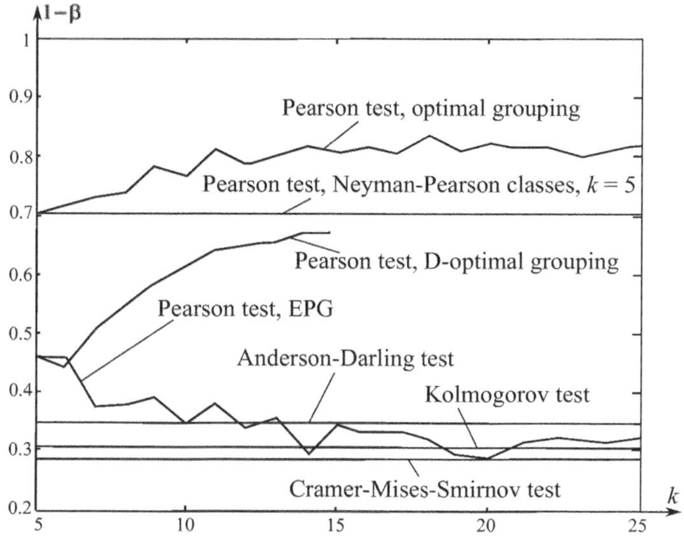

Figure 3.1. *The power of the χ^2 Pearson test when testing the simple hypothesis*

Figures 3.1 and 3.2 show the power of the Pearson χ^2 test for the hypotheses H_0 and H_3 of our example in the case of different grouping methods, depending on the number of intervals ($\alpha = 0.1, n = 500$). The powers of the Kolmogorov, Cramer-von Mises-Smirnov and Anderson-Darling goodness-of-fit tests are given for comparison.

3.3. Comparison of goodness-of-fit tests for censored data

In [LEM 10a], we investigated the power of the modified Kolmogorov, Cramer-von Mises-Smirnov and Anderson-Darling tests in the case of type II censored samples, and in [CHI 11c], these tests were compared with the NRR χ^2 test by power. The power of classical Kolmogorov, Cramer-von Mises-Smirnov and Anderson-Darling tests for samples completed from original randomly censored samples was studied in [CHI 11b]. It was shown that the power of the tests considered increases as the sample size grows. It is necessary to note that in the case of type I and random censoring we cannot simulate the same censoring scheme for both hypotheses H_0 and H_1, thus it is not correct to compare different tests by power in such situations.

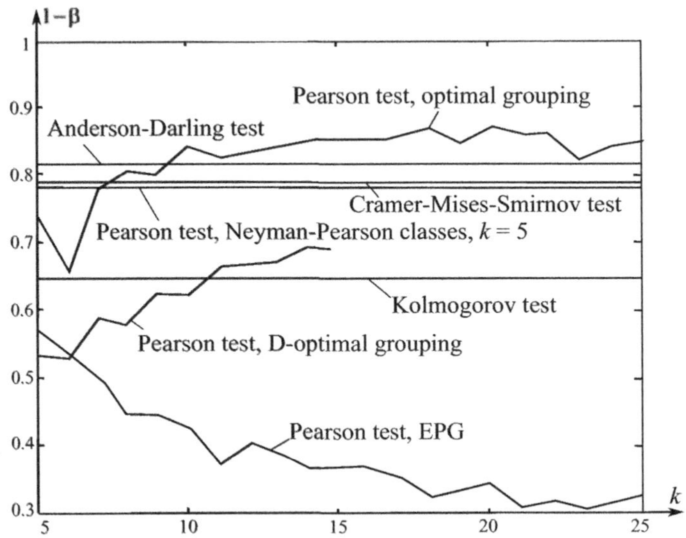

Figure 3.2. *The power of the χ^2 Pearson test when testing the composite hypothesis*

In this section, we compare the goodness-of-fit tests by power based on the example of four different pairs of competing hypotheses by means of computer simulation methods. The estimates of the power are calculated based on the distributions of the test statistics $G(s|H_0)$ and $G(s|H_1)$, which are simulated for type II censored samples of size $n = 200$. In the case of composite hypotheses, the distribution parameters are estimated by the maximum likelihood method. The number of simulations is $N = 10^5$. The values of the power of the tests are calculated for the significance level $\alpha = 0.1$.

3.3.1. *Lognormal-generalized Weibull pair of competing hypotheses*

The null hypothesis H_0 is the lognormal distribution with the distribution function $F(t;\theta,\nu) = \frac{1}{2} - \frac{1}{2\sqrt{\pi}}\Gamma\left(\frac{1}{2\nu^2}\ln^2\left(\frac{t}{\theta}\right), \frac{1}{2}\right)$ and the parameters $\theta = 2$, $\nu = 2$. The competing hypothesis H_1 is the generalized Weibull distribution with the distribution function $F(t;\theta,\nu,\lambda) = 1 - \exp\left(1 - \left(1 + \left(\frac{t}{\theta}\right)^{\nu}\right)^{1/\lambda}\right)$ and the parameters $\theta = 0.35$, $\nu = 1$, $\lambda = 3.36$. The distribution functions of these models do not intersect, and the hazard rate functions of both models decrease.

In Table 3.4, the powers of the tests considered are given for various censoring degrees in the case of testing the simple hypothesis. The modified Kolmogorov, Cramer-von Mises-Smirnov and Anderson-Darling tests are denoted as S_K^M, S_ω^M and S_Ω^M, respectively. The classical Kolmogorov, Cramer-von Mises-Smirnov and Anderson-Darling tests for completed samples are denoted as S_K^R, S_ω^R and S_Ω^R, respectively. In the case of testing composite hypotheses, the NRR χ^2 test is denoted as Y^2 with the number of intervals k.

	Censoring degrees								
Test	0%	5%	10%	20%	30%	40%	50%	60%	70%
S_K^M	0.29	0.28	0.29	0.26	0.24	0.21	0.17	0.14	0.11
S_ω^M	0.32	0.31	0.30	0.25	0.22	0.17	0.14	0.11	0.10
S_Ω^M	0.35	0.32	0.30	0.25	0.22	0.19	0.17	0.15	0.14
S_K^R	0.29	0.28	0.29	0.27	0.26	0.23	0.19	0.15	0.12
S_ω^R	0.32	0.32	0.32	0.29	0.27	0.24	0.20	0.16	0.13
S_Ω^R	0.35	0.34	0.33	0.30	0.27	0.23	0.20	0.17	0.14
Y^2,3	0.29	0.24	0.21	0.18	0.15	0.13	0.11	0.10	0.10
Y^2,5	0.18	0.15	0.14	0.11	0.10	0.10	0.10	0.10	0.10

Table 3.4. *Test power for the lognormal-generalized Weibull pair when testing the simple hypothesis*

As you can see from Table 3.4, the power estimates of the tests considered decrease as the censoring degree grows. The classical tests for completed samples have the higher power than the corresponding modified tests, and the Anderson-Darling is the most powerful among the tests considered. For this pair of competing hypotheses, the power of the NRR χ^2 test turned out to be the lowest and for large censoring degrees, it is even equal to the significance level.

In Table 3.5, the powers of the tests considered are given in the case of testing the composite hypothesis.

	Censoring degrees								
Test	0%	5%	10%	20%	30%	40%	50%	60%	70%
S_K^M	0.15	0.15	0.15	0.15	0.16	0.16	0.16	0.17	0.17
S_ω^M	0.16	0.16	0.16	0.16	0.17	0.17	0.18	0.18	0.18
S_Ω^M	0.19	0.19	0.19	0.19	0.20	0.20	0.21	0.21	0.21
S_K^R	0.15	0.15	0.15	0.15	0.15	0.14	0.14	0.12	0.11
S_ω^R	0.16	0.17	0.17	0.16	0.17	0.16	0.15	0.13	0.12
S_Ω^R	0.19	0.19	0.19	0.19	0.19	0.18	0.17	0.15	0.13
Y^2,3	0.17	0.15	0.15	0.15	0.15	0.15	0.15	0.15	0.15
Y^2,5	0.15	0.14	0.14	0.13	0.14	0.13	0.13	0.13	0.13

Table 3.5. *Test power for the lognormal-generalized Weibull pair when testing the composite hypothesis*

As can be seen from Table 3.5, the modified Anderson-Darling test turns out to be the most powerful in comparison with the Cramer-von Mises-Smirnov, Kolmogorov and Nikulin-Rao-Robson tests. For censoring degrees up to 20%, the power of the modified and classical Anderson-Darling tests is almost equal, but beginning from 20% the power of the classical tests for completed samples decreases, whereas the power of the modified Kolmogorov, Cramer-von Mises-Smirnov and Anderson-Darling tests increases at the same situation. The power of the NRR χ^2 test practically does not change with the growth of the censoring degree. It is important to note that the power of the NRR χ^2 test is less in the case of $k = 5$ than in the case of $k = 3$.

3.3.2. *Exponential-Weibull pair of competing hypotheses*

The null hypothesis H_0 is the exponential distribution with the distribution function $F(t;\theta) = 1 - \exp\left(-\frac{t}{\theta}\right)$ and the parameter $\theta = 1$. The competing hypothesis H_1 is the Weibull distribution with the parameters $\theta = 1$, $\nu = 1.2$. The distribution functions of these models intersect once, the exponential distribution has a constant hazard rate and the hazard rate function of the Weibull distribution increases.

In Table 3.6, the powers of the tests considered are given for various censoring degrees in the case of testing the simple hypothesis.

	Censoring degrees								
Test	0%	5%	10%	20%	30%	40%	50%	60%	70%
S_K^M	0.38	0.39	0.37	0.36	0.35	0.37	0.42	0.49	0.58
S_ω^M	0.42	0.40	0.37	0.35	0.37	0.43	0.50	0.59	0.67
S_Ω^M	0.56	0.49	0.44	0.42	0.45	0.51	0.56	0.62	0.68
S_K^R	0.38	0.39	0.38	0.36	0.35	0.34	0.34	0.35	0.37
S_ω^R	0.42	0.43	0.40	0.36	0.33	0.33	0.35	0.38	0.40
S_Ω^R	0.56	0.55	0.50	0.43	0.39	0.39	0.41	0.43	0.45
Y^2,3	0.76	0.70	0.65	0.58	0.53	0.52	0.52	0.53	0.57
Y^2,5	0.72	0.66	0.61	0.55	0.52	0.50	0.49	0.51	0.53

Table 3.6. *Test power for the exponential-Weibull pair when testing the simple hypothesis*

In contrast to the previous pair of competing hypotheses in the case of testing the simple hypothesis, the power changes non-monotonically. The power of the tests decreases, but beginning with some censoring degree, it grows. Moreover, for large censoring degrees, the power of the modified tests turns out to be even larger than the power of these tests in the case of complete samples. Among the tests considered, the most powerful one is the NRR χ^2 test, but for large censoring degrees (more than 50%), it yields to the modified Anderson-Darling and Cramer-von Mises-Smirnov tests.

In Table 3.7, the powers of the tests considered are given in the case of testing the composite hypothesis.

	Censoring degrees								
Test	0%	5%	10%	20%	30%	40%	50%	60%	70%
S_K^M	0.84	0.80	0.76	0.69	0.62	0.55	0.47	0.40	0.33
S_ω^M	0.90	0.87	0.83	0.76	0.69	0.61	0.53	0.45	0.38
S_Ω^M	0.91	0.88	0.85	0.79	0.72	0.65	0.56	0.47	0.38
S_K^R	0.84	0.79	0.73	0.63	0.54	0.45	0.35	0.26	0.16
S_ω^R	0.90	0.86	0.81	0.71	0.61	0.50	0.38	0.27	0.17
S_Ω^R	0.91	0.88	0.84	0.74	0.65	0.55	0.45	0.35	0.24
Y^2,3	0.83	0.78	0.74	0.66	0.59	0.51	0.44	0.37	0.30
Y^2,5	0.79	0.74	0.69	0.62	0.54	0.47	0.39	0.33	0.27

Table 3.7. *Test power for the exponential-Weibull pair when testing the composite hypothesis*

When the composite hypothesis is tested, the power of all the tests considered decreases with the increasing number of censored observations. The modified Anderson-Darling test has the higher power than the other non-parametric tests. It is interesting to note that when comparing the classical tests for completed samples, the Anderson-Darling test is again found to be most preferable. The power estimates of the NRR χ^2 test are higher for the lower number of intervals ($k = 3$).

3.3.3. *Weibull-generalized Weibull pairs of competing hypotheses*

Let us consider two pairs of competing hypotheses. The null hypothesis H_0 is the Weibull distribution with the parameters $\theta = 2$, $\nu = 2$. In the case of the first pair, the competing hypothesis H_1 is the generalized Weibull distribution with the parameters $\theta = 2$, $\nu = 2$, $\lambda = 0.9$. In the case of the second pair, the competing hypothesis H_1 is the generalized Weibull distribution with the parameters $\theta = 2$, $\nu = 2$, $\lambda = 1.1$.

In Tables 3.8 and 3.9, the powers of the tests considered are given for various censoring degrees when testing the simple hypothesis.

	Censoring degrees								
Test	0%	5%	10%	20%	30%	40%	50%	60%	70%
S_K^M	0.49	0.49	0.49	0.45	0.40	0.34	0.27	0.22	0.16
S_ω^M	0.55	0.53	0.50	0.43	0.35	0.29	0.22	0.16	0.11
S_Ω^M	0.58	0.53	0.49	0.41	0.33	0.28	0.22	0.18	0.14
S_K^R	0.49	0.49	0.49	0.47	0.43	0.37	0.30	0.23	0.16
S_ω^R	0.55	0.54	0.54	0.50	0.44	0.38	0.31	0.24	0.17
S_Ω^R	0.58	0.57	0.55	0.49	0.43	0.36	0.29	0.23	0.18
Y^2,3	0.51	0.45	0.41	0.32	0.26	0.20	0.15	0.12	0.08
Y^2,5	0.35	0.29	0.26	0.19	0.15	0.11	0.09	0.07	0.05

Table 3.8. *Test power for the first Weibull-generalized Weibull pair when testing the simple hypothesis*

As we can see from Table 3.8, for the first pair of competing hypotheses, when the generalized parameter is equal to 0.9, the most powerful test is the classical Anderson-Darling test for samples completed from the original censored samples. The power of the NRR χ^2 test is very low in this case. Comparing the modified non-parametric tests, we can see that for the censoring degrees larger than 20%, the Kolmogorov test has higher power than the modified Anderson-Darling and Cramer-von Mises-Smirnov tests, which was not observed for the other pairs of competing hypotheses.

Test	Censoring degrees								
	0%	5%	10%	20%	30%	40%	50%	60%	70%
S_K^M	0.43	0.43	0.43	0.43	0.41	0.38	0.34	0.30	0.26
S_ω^M	0.48	0.48	0.48	0.46	0.42	0.39	0.35	0.31	0.28
S_Ω^M	0.56	0.55	0.53	0.47	0.41	0.36	0.31	0.27	0.23
S_K^R	0.43	0.43	0.43	0.42	0.40	0.37	0.32	0.26	0.20
S_ω^R	0.48	0.47	0.47	0.46	0.42	0.38	0.32	0.26	0.20
S_Ω^R	0.56	0.55	0.53	0.48	0.43	0.37	0.31	0.25	0.20
Y^2,3	0.64	0.59	0.55	0.48	0.42	0.37	0.31	0.27	0.23
Y^2,5	0.60	0.55	0.52	0.44	0.39	0.34	0.29	0.26	0.21

Table 3.9. *Test power for the second Weibull-generalized Weibull pair when testing the simple hypothesis*

As can be seen from Table 3.9, when the generalized parameter is equal to 1.1 and the censoring degree is less than 30%, the most powerful test is the NRR χ^2 test with the number of intervals $k = 3$. For the censoring degrees larger than 30%, this test yields by power to the other non-parametric tests. In the case of small censoring degrees, the modified Kolmogorov, Cramer-von Mises-Smirnov and Anderson-Darling tests coincide by power with the corresponding classical tests for completed samples. However, for large censoring degrees, the modified non-parametric tests become more powerful than the other tests.

The estimates of the power of all the tests considered in the case of the composite hypotheses turned out to be equal to the significance level ($\alpha = 0.1$), that is, none of the tests considered can distinguish between the Weibull distribution and the generalized Weibull distribution, which can be explained as follows: for the generalized

Weibull distribution with the chosen parameter values, it is possible to fit a very close Weibull distribution.

So, what is the answer to the question: "Which goodness-of-fit test is most preferable for right censored samples"? There is no unique test which could be the best in any situation. Nevertheless, the results presented allow us to formulate some advantages and disadvantages of the tests considered. Concerning the modified Kolmogorov, Cramer-von Mises-Smirnov and Anderson-Darling tests, in the case of type I and type II censored data, we can use the tables of percentage points for these tests in simple hypothesis testing [LEM 10a]. While testing a composite hypothesis, it is possible to simulate the distribution of test statistics for the distribution under test and censoring scheme given. The modified Anderson-Darling test was the most powerful in most cases.

There is a good possibility to use the transformation of a censored sample to a complete one and then to apply the classical Kolmogorov, Cramer-von Mises-Smirnov and Anderson-Darling tests for a completed sample. However, this approach can be used only when testing simple hypotheses, or in the case of small censoring degrees, when testing composite hypotheses. These tests almost do not yield by power to the corresponding modified tests.

The NRR test has a number of advantages compared with the tests considered. In particular, there is no significant dependence of the NRR statistic distributions (in the case of limited sample sizes) on the distribution of censoring times. As the sample size grows, NRR statistic distributions converge to the corresponding χ^2 distribution law. In this section, we considered this test as a non-parametric test, but for some given pairs of competing hypotheses, it is possible to find the optimal number of grouping intervals and boundary points that maximize the test power.

4

Chi-squared Goodness-of-fit Tests for Regression Models

Suppose that n independent failure time variables are observed. Let us consider the hypothesis H_0 stating that the survival function given the vector of explanatory variables (covariates)

$$z(t) = (z_0(t), z_1(t), ..., z_m(t))^T, \quad z_0(t) \equiv 1,$$

has the form:

$$S(t|z) = S0(t; \theta, z), \qquad [4.1]$$

where S_0 is a specified functional of time t, finite-dimensional parameter $\theta = (\theta_0, \theta_1, ..., \theta_s)^T \in \Theta \subset R^{s+1}$ and z. The hypothesis H_0 can be also formulated in terms of the hazard functions $\lambda(t|z) = -S'(t|z)/S(t|z)$ or the cumulative hazard functions $\Lambda(t|z) = -\ln S(t|z)$.

Let us consider the examples of such hypotheses (see [AAL 78], [BAG 02], [BAG 10b], [BIL 68], [LEC 83]):

1) Parametric accelerated failure time (AFT) model: $\theta = (\beta^T, \gamma^T)^T$,

$$S(t|z) = S_0\left(\int_0^t e^{-\beta^T z(u)} du; \gamma\right), \qquad [4.2]$$

where $\beta = (\beta_0, \cdots, \beta_m)^T$ is a vector of unknown regression parameters, the function S_0 does not depend on z_i and belongs to a specified class of survival functions: $S_0(t, \gamma),\ \gamma = (\gamma_1, ..., \gamma_q)^T \in G \subset R^q$. If explanatory variables are constant over time, then the parametric AFT model has the form $S(t|z) = S_0\left(e^{-\beta^T z} t; \gamma\right)$, and the logarithm of the failure time T under z may be written as:

$$\ln\{T\} = \beta^T z + \varepsilon, \quad \varepsilon \sim S(t) = S_0(\ln t).$$

If ε is a normally distributed random variable, then the AFT model is a standard multiple linear regression model.

Sometimes some specified functions of z instead of z may be used. In accelerated life testing transforms, $\ln z$ (power rule model), $1/z$ (Arrhenius model), $\ln \frac{z}{1-z}$ (Meeker–Luvalle model) and others are used.

2) Parametric proportional hazards (Cox) model: $\theta = (\beta^T, \gamma^T)^T$,

$$\lambda(t|z) = e^{\beta^T z(t)} \lambda_0(t, \gamma), \qquad [4.3]$$

where λ_0 belongs to a specified class of hazard functions $\lambda(t, \gamma),\ \gamma \in G \subset R^q$.

3) Parametric generalized proportional hazards models (including parametric frailty and linear transformations models): $\theta = (\beta^T, \gamma^T, \nu^T)^T$,

$$h(\Lambda(t|z), \nu) = \int_0^t e^{\beta^T z(u)} \lambda_0(u, \gamma) du, \qquad [4.4]$$

where the function $h(x, \nu)$ and the hazard function $\lambda_0(t, \gamma)$ belong to a specified parametric classes. In particular, if $h(x, \nu) = [(1+x)^\nu - 1]/\nu$, $h(x, \nu) = [1 - e^{-\nu x}]/\nu$ and $h(x, \nu) = x + \nu x^2/2$, we have respectively parametric positive stable, gamma, and inverse Gaussian frailty models with explanatory variables.

4) Models with cross effects of survival functions:

$$\lambda(t|z) = g(z, \beta, \nu, \Lambda_0(t, \gamma)), \qquad [4.5]$$

where the cumulative hazard Λ_0 has a specified form and the function g has one of the following forms:

$$g(z,\beta,\nu,x) = e^{\beta^T z}\left[1 + e^{(\beta+\nu)^T z}x\right]^{e^{-\nu^T z}-1}$$

$$\frac{e^{\beta^T z + xe^{\nu^T z}}}{1 + e^{(\beta+\nu)^T z}[e^{xe^{\nu^T z}} - 1]}. \qquad [4.6]$$

We give chi-squared type goodness-of-fit tests for general hypothesis H_0 with possibly time-dependent covariates. The choice of random grouping intervals as data functions is considered.

Simple formulas useful for computing of test statistics are derived.

4.1. Data and the idea of chi-squared test construction

We shall give chi-squared tests for testing the hypothesis H_0 from right censored failure time regression data:

$$(X_1,\delta_1,z_1),\cdots,(X_n,\delta_n,z_n),$$

where $X_i = T_i \wedge C_i$, $\delta_i = \mathbf{1}_{\{T_i \le C_i\}}$, T_i is the failure time, C_i is the censoring time, $i = 1, ..., n$ and the vector of covariates is denoted as:

$$z_i(t) = (z_{i0}(t), z_{i1}(t), ..., z_{im}(t))^T, \quad z_{i0}(t) \equiv 1.$$

Set

$$N_i(t) = \mathbf{1}_{\{X_i \le t, \delta_i = 1\}}, \quad Y_i(t) = \mathbf{1}_{\{X_i \ge t\}},$$

$$N(t) = \sum_{i=1}^n N_i(t), \quad Y(t) = \sum_{i=1}^n Y_i(t).$$

Suppose that:

1) the processes N_i, Y_i, z_i are observed up to the finite time τ;

2) survival distributions of all n objects, given z_i, are absolutely continuous with the survival functions $S_i(t)$ and the hazard rates $\lambda_i(t)$;

3) censoring is non-informative and the multiplicative intensity model holds: the compensators of the counting processes N_i with respect to the history of the observed processes are $\int Y_i \lambda_i du$.

Divide the interval $[0, \tau]$ into k smaller intervals $I_j = (a_{j-1}, a_j]$. $a_0 = 0$, $a_k = \tau$, and denote by

$$U_j = N(a_j) - N(a_{j-1})$$

the number of observed failures in the j-th interval, $j = 1, 2...., k$.

What is the "dynamic conditional expected" number of observed failures in the interval I_j under H_0 given the data? Set

$$\lambda_i(t, \theta) := \lambda(t; \theta, z_i).$$

Under regularity conditions, the equality

$$EN_i(t) = E \int_0^t \lambda_i(u, \theta) Y_i(u) du \qquad [4.7]$$

holds; here $\theta = (\beta^T, \gamma^T)^T$ and $\lambda_i(t, \theta)$ is the hazard function of T_i under z_i.

The equality [4.7] suggests that we can "conditionally expect" to observe

$$e_j = \sum_{i=1}^n \int_{I_j} \lambda_i(u, \hat{\theta}) Y_i(u) du \qquad [4.8]$$

failures in the interval I_j; here $\hat{\theta}$ is the ML estimator of θ under H_0.

So, a test can be based on the vector

$$Z = (Z_1, ..., Z_k)^T, \quad Z_j = \frac{1}{\sqrt{n}}(U_j - e_j), \quad j = 1, ..., k, \qquad [4.9]$$

of differences between the number of observed and "expected" failures in the intervals $I_1, ..., I_k$.

4.2. Asymptotic distribution of the random vector Z

To investigate the properties of the statistic Z, we need properties of the stochastic process:

$$H_n(t) = \frac{1}{\sqrt{n}}(N(t) - \sum_{i=1}^{n} \int_0^t \lambda_i(u, \hat{\theta}) Y_i(u) du).$$

To obtain these properties, we use the properties of the ML estimators which are well known.

Conditions A: $\hat{\theta} \xrightarrow{P} \theta_0$, $\frac{1}{\sqrt{n}}\dot{\ell}(\theta_0) \xrightarrow{d} N_m(0, \mathbf{i}(\theta_0))$, $-\frac{1}{n}\ddot{\ell}(\theta_0) \xrightarrow{P} \mathbf{i}(\theta_0)$ and $\sqrt{n}(\hat{\theta} - \theta_0) = \mathbf{i}^{-1}(\theta_0)\ \frac{1}{\sqrt{n}}\dot{\ell}(\theta_0) + o_P(1)$, where

$$\dot{\ell}(\theta) = \sum_{i=1}^{n} \int_0^{\tau} \frac{\partial}{\partial \theta} \ln \lambda_i(u, \theta) \left\{ dN_i(u) - Y_i(u)\lambda_i(u, \theta) du \right\}$$

is the score function.

Conditions A (they mean consistency and asymptotic normality of the ML estimator $\hat{\theta}$) hold, in particular, if conditions VI.1.1 given in [AND 93] hold. Set

$$S^{(0)}(t, \theta) = \sum_{i=1}^{n} Y_i(t)\lambda_i(t, \theta),$$

$$S^{(1)}(t, \theta) = \sum_{i=1}^{n} Y_i(t) \frac{\partial \ln \lambda_i(t, \theta)}{\partial \theta} \lambda_i(t, \theta),$$

$$S^{(2)}(t, \theta) = \sum_{i=1}^{n} Y_i(t) \frac{\partial^2 \ln \lambda_i(t, \theta)}{\partial \theta^2} \lambda_i(t, \theta).$$

Conditions B: There exist a neighborhood Θ_0 of θ_0 and continuous bounded on $\Theta_0 \times [0, \tau]$ functions

$$s^{(0)}(t, \theta), \quad s^{(1)}(t, \theta) = \frac{\partial s^{(0)}(t, \theta)}{\partial \theta}, \quad s^{(2)}(t, \theta) = \frac{\partial^2 s^{(0)}(t, \theta)}{\partial \theta^2},$$

such that, for $j = 0, 1, 2$,

$$\sup_{t \in [0,\tau], \theta \in \Theta_0} \|\frac{1}{n} S^{(j)}(t, \theta) - s^{(j)}(t, \theta)\| \xrightarrow{P} 0 \quad \text{as} \quad n \to \infty.$$

Conditions B imply that uniformly for $t \in [0, \tau]$,

$$\frac{1}{n}\sum_{i=1}^{n}\int_0^t \lambda_i(u, \theta_0)Y_i(u)du \xrightarrow{P} A(t),$$

$$\frac{1}{n}\sum_{i=1}^{n}\int_0^t \frac{\partial}{\partial\theta}\lambda_i(u, \theta_0)Y_i(u)du \xrightarrow{P} C(t),$$

where A and C are finite functions.

LEMMA 4.1.– *Under conditions A and B, the following convergence holds:*

$$H_n \xrightarrow{d} V \quad \text{on } D[0, \tau];$$

here $D[0, \tau]$ *is the space of cadlag functions with Skorokhod metric and* V *is a zero-mean Gaussian martingale such that, for all* $0 \leq u \leq v \leq \tau$,

$$cov(V(u), V(v)) = A(u) - C^T(u)\mathbf{i}^{-1}(\theta_0)C(v).$$

PROOF–. By conditions A

$$\sqrt{n}(\hat{\theta} - \theta_0) = \mathbf{i}^{-1}(\theta_0)\,\frac{1}{\sqrt{n}}\sum_{i=1}^{n}\int_0^\tau \frac{\partial}{\partial\theta}\ln\lambda_i(u, \theta_0)\,dM_i(u) + o_P(1),$$

$$dM_i(u) = dN - Y_i(u)\lambda_i(u, \theta_0)du.$$

Using conditions B and the Taylor formula, we have:

$$H_n(t) = \frac{1}{\sqrt{n}}(N(t) - \sum_{i=1}^{n}\int_0^t \lambda_i(u, \hat{\theta})Y_i(u)du)$$

$$= \frac{1}{\sqrt{n}}M(t) - \sqrt{n}\int_0^t [s^{(0)}(u, \hat{\theta}) - s^{(0)}(u, \theta_0)]du + o_P(1)$$

$$= \frac{1}{\sqrt{n}}M(t) - \int_0^t [s^{(1)}(u, \theta_0)]^T du\sqrt{n}(\hat{\theta} - \theta_0) + o_P(1)$$

$$= \frac{1}{\sqrt{n}}M(t) - C^T(t)\,\mathbf{i}^{-1}(\theta_0)\frac{1}{\sqrt{n}}\sum_{i=1}^{n}\int_0^\tau \frac{\partial}{\partial\theta}\ln\lambda_i(u, \theta_0)\,dM_i(u)$$

$$+ o_P(1)$$

$$=: M_1^*(t) - C^T(t)\,\mathbf{i}^{-1}(\theta_0)M_2^*(\tau).$$

The predictable variations and covariations are:

$$< M_1^* > (t) = \frac{1}{n}\int_0^t S^{(0)}(u,\theta)du \xrightarrow{P} A(t),$$

$$< M_1^*, M_2^* > (t) = \frac{1}{n}\sum_{i=1}^n \int_0^t \dot{\lambda}_i(u,\theta_0)Y_i(u)du \xrightarrow{P} C(t),$$

$$< M_2^* > (\tau) = \frac{1}{n}\sum_{i=1}^n \int_0^\tau \frac{\partial}{\partial\theta}\ln\lambda_i(u,\theta)$$

$$(\frac{\partial}{\partial\theta}\ln\lambda_i(u,\theta))^T\lambda_i(u,\theta_0)Y_i(u)du \xrightarrow{P} \mathbf{i}(\theta_0).$$

The result of the lemma is implied by the CLT for martingales because conditions A and B imply that the Lindeberg condition (see [AND 93]) is obviously satisfied: for all $t \in [0,\tau]$,

$$| < \frac{1}{\sqrt{n}}\int_0^t \mathbf{1}_{\{\frac{1}{\sqrt{n}}>\varepsilon\}}\, dM(u) > |$$

$$= \frac{1}{n}\mathbf{1}_{\{\frac{1}{\sqrt{n}}>\varepsilon\}}\sum_{i=1}^n\int_0^t \lambda_i(u,\theta_0)Y_i(u)du \to 0,$$

$$| < \frac{1}{\sqrt{n}}\sum_{i=1}^n\int_0^\tau \frac{\partial}{\partial\theta_j}\ln\lambda_i(u,\theta_0)\mathbf{1}_{\{\frac{1}{\sqrt{n}}\frac{\partial}{\partial\theta_j}\ln\lambda_i(u,\theta_0)>\varepsilon\}}\, dM_i(u) > |$$

$$= \frac{1}{n}\sum_{i=1}^n\int_0^\tau \{\frac{\partial}{\partial\theta_j}\ln\lambda_i(u,\theta_0)\}^2\mathbf{1}_{\{\frac{1}{\sqrt{n}}\frac{\partial}{\partial\theta_j}\ln\lambda_i(u,\theta_0)>\varepsilon\}}$$

$$\lambda_i(u,\theta_0)Y_i(u)du \to 0.$$

The proof is complete.

For $i = 0, ..., s;\ j, j' = 1, ..., k$, set

$$V_j = V(a_j) - V(a_{j-1}), \quad \sigma_{jj'} = cov(V_j, V_{j'}),$$

$$A_j = A(a_j) - A(a_{j-1}), \quad C_{ij} = C_i(a_j) - C_i(a_{j-1}),$$

$$\mathcal{C}_j = (C_{0j}, ..., C_{sj})^T,$$
$$\Sigma = [\sigma_{jj'}]_{k\times k}, \quad \mathcal{C} = [C_{ij}]_{(s+1)\times k},$$

and denote by $\mathcal{A}$ a $k \times k$ diagonal matrix with the diagonal elements $A_1, ..., A_k$.

THEOREM 4.1.– *Under conditions A and B*

$$Z \xrightarrow{d} Y \sim N_k(0, \Sigma) \quad as \ n \to \infty, \qquad [4.10]$$

where

$$\Sigma = \mathcal{A} - \mathcal{C}^T \mathbf{i}^{-1}(\theta_0)\mathcal{C}. \qquad [4.11]$$

PROOF.– Set $\mathbf{i} = ||i_{ll'}|| = \mathbf{i}(\theta_0)$. Lemma 1 implies that:

$$\begin{aligned}\sigma_{jj} &= Var(V(a_j) - V(a_{j-1})) = A(a_j) - C^T(a_j)\mathbf{i}^{-1}C(a_j) \\ &\quad - 2[A(a_{j-1}) - C^T(a_{j-1})\mathbf{i}^{-1}C(a_j)] \\ &\quad + A(a_{j-1}) - C^T(a_j)\mathbf{i}^{-1}C(a_{j-1}) \\ &= A_j - \mathcal{C}_j^T\mathbf{i}^{-1}C(a_j) + C^T(a_{j-1})\mathbf{i}^{-1}\mathcal{C}_j \\ &= A_j - C^T(a_j)\mathbf{i}^{-1}\mathcal{C}_j + C^T(a_{j-1})\mathbf{i}^{-1}\mathcal{C}_j \\ &= A_j - \mathcal{C}_j^T\mathbf{i}^{-1}(\theta_0)\mathcal{C}_j.\end{aligned}$$

For $j < j'$,

$$\begin{aligned}\sigma_{jj'} &= cov(V(a_j) - V(a_{j-1}), V(a_{j'}) - V(a_{j'-1})) \\ &= A(a_j) - C^T(a_j)\mathbf{i}^{-1}C(a_{j'}) - A(a_j) \\ &\quad + C^T(a_j)\mathbf{i}^{-1}C(a_{j'-1}) - A(a_{j-1}) + C^T(a_{j-1})\mathbf{i}^{-1}C(a_{j'}) \\ &\quad + A(a_{j-1}) - C^T(a_{j-1})\mathbf{i}^{-1}C(a_{j'-1}) \\ &= -\mathcal{C}_j^T\mathbf{i}^{-1}C(a_{j'}) + \mathcal{C}_j^T\mathbf{i}^{-1}C(a_{j'-1}) = -\mathcal{C}_j^T\mathbf{i}^{-1}(\theta_0)\mathcal{C}_{j'}.\end{aligned}$$

The proof is complete.

Set $G = \mathbf{i} - CA^{-1}C^T$. The formula

$$\Sigma^- = \mathcal{A}^{-1} + \mathcal{A}^{-1}\mathcal{C}^T G^- \mathcal{C}\mathcal{A}^{-1}$$

implies that we need to invert only the diagonal $k \times k$ matrix A and find the general inverse of the $s \times s$ matrix G.

THEOREM 4.2.– *Under conditions A and B, the following estimators of A_j, C_j, $\mathbf{i}(\theta_0)$ and Σ are consistent:*

$$\hat{A}_j = U_j/n, \quad \hat{\mathcal{C}}_j = \frac{1}{n}\sum_{i=1}^n \int_{I_j} \frac{\partial}{\partial\theta} \ln \lambda_i(u, \hat{\theta}) dN_i(u), \qquad [4.12]$$

and

$$\hat{\mathbf{i}} = \frac{1}{n}\sum_{i=1}^n \int_0^T \frac{\partial \ln \lambda_i(u,\hat{\theta})}{\partial\theta} \left(\frac{\partial \ln \lambda_i(u,\hat{\theta})}{\partial\theta}\right)^T dN_i(u), \qquad [4.13]$$

$$\hat{\Sigma} = \hat{\mathcal{A}} - \hat{\mathcal{C}}^T \hat{\mathbf{i}}^{-1} \hat{\mathcal{C}}; \qquad [4.14]$$

PROOF.– Similarly as in the proof of Theorem 4.1, we have:

$$\hat{A}_j - A_j = \frac{1}{n}\sum_{i=1}^n \int_{I_j} dN_i(u) - A_j$$

$$= \frac{1}{n}\sum_{i=1}^n \int_{I_j} dM_i(u) + \frac{1}{n}\sum_{i=1}^n \int_{I_j} \lambda_i(u,\theta_0) Y_i(u) du - A_j \xrightarrow{P} 0,$$

$$\hat{\mathcal{C}}_j - \mathcal{C}_j = \frac{1}{n}\sum_{i=1}^n \int_{I_j} \frac{\partial}{\partial\theta} \ln \lambda_i(u,\hat{\theta}) dM_i(u)$$

$$+ \frac{1}{n}\sum_{i=1}^n \int_{I_j} \left(\frac{\partial}{\partial\theta} \ln \lambda_i(u,\hat{\theta}) - \frac{\partial}{\partial\theta} \ln \lambda_i(u,\theta_0)\right) \lambda_i(u,\theta_0) du$$

$$+ \frac{1}{n}\sum_{i=1}^n \int_{I_j} \frac{\partial}{\partial\theta} \ln \lambda_i(u,\theta_0) \lambda_i(u,\theta_0) du - \mathcal{C}_j \xrightarrow{P} 0.$$

By conditions A:

$$\hat{\mathbf{i}} = -\frac{1}{n}\ddot{\ell}(\hat{\theta}) + o_P(1) \xrightarrow{P} \mathbf{i}(\theta_0).$$

4.3. Test statistic

Theorems 4.1 and 4.2 imply that a test for the hypothesis H_0 can be based on the statistic $Y^2 = Z^T\hat{\Sigma}^- Z$, where $\hat{\Sigma}^- = \hat{A}^{-1} + \hat{C}^{-1}\hat{A}^T\hat{G}^-\hat{C}\hat{A}^{-1}$, $\hat{G} = \hat{\mathbf{i}} - \hat{C}\hat{A}^{-1}\hat{C}^T$. This statistic can be written in the form:

$$Y_n^2 = \sum_{j=1}^{k} \frac{(U_j - e_j)^2}{U_j} + Q, \qquad [4.15]$$

where

$$U_j = \sum_{i:X_i \in I_j} \delta_i, \quad e_j = \sum_{i=1}^{n} \int_{I_j} \lambda_i(u, \hat{\theta}) Y_i(u) du,$$

$$Q = W^T \hat{G}^- W,$$

$$W = (W_0, ..., W_s)^T, \quad \hat{G} = [\hat{g}_{ll'}]_{s \times s},$$

$$\hat{g}_{ll'} = \hat{i}_{ll'} - \sum_{j=1}^{k} \hat{C}_{lj} \hat{C}_{l'j} \hat{A}_j^{-1}, \qquad [4.16]$$

$$\hat{C}_{lj} = \frac{1}{n} \sum_{i:X_i \in I_j} \delta_i \frac{\partial}{\partial \theta_l} \ln \lambda_i(X_i, \hat{\theta}),$$

$$\hat{i}_{ll'} = \frac{1}{n} \sum_{i=1}^{n} \delta_i \frac{\partial \ln \lambda_i(X_i, \hat{\theta})}{\partial \theta_l} \frac{\partial \ln \lambda_i(X_i, \hat{\theta})}{\partial \theta_{l'}}, \qquad [4.17]$$

$$W_l = \sum_{j=1}^{k} \hat{C}_{lj} \hat{A}_j^{-1} Z_j, \quad l, l' = 0, ..., s. \qquad [4.18]$$

The limit distribution of the statistic Y_n^2 is chi-square with $r = rank(V^-) = Tr(V^- V)$ degrees of freedom. If the matrix G is non-degenerate, then $r = k$.

Test for the hypothesis H_0**:** the hypothesis is rejected with the approximate significance level α if $Y_n^2 > \chi^2_\alpha(r)$.

4.4. Choice of random grouping intervals

A usual experimental plan in accelerated life testing is to test several groups of units under different higher stress conditions. In such an experiment, it is possible that the failures of units from different groups are mostly concentrated in different non-intersecting intervals. So, using the common idea of constructing chi-square test by dividing the interval $[0, \tau)$ into smaller intervals and comparing the observed and expected numbers of failures, the choice of the ends of the intervals is very important because dividing the intervals into equal lengths may give intervals in which the number of observed failures is zero or very small.

Let us consider the choice of the limits of grouping intervals as random data functions.

Define:

$$E_k = \sum_{i=1}^{n} \int_0^{\tau} \lambda_i(u, \hat{\theta}) Y_i(u) du = \sum_{i=1}^{n} \Lambda_i(X_i, \hat{\theta}),$$

$$E_j = \frac{j}{k} E_k, \quad j = 1, ..., k. \qquad [4.19]$$

So, we seek $\hat{a}_j$ to have equal numbers of expected failures (not necessary an integer) in all intervals. So, $\hat{a}_j$ satisfy the equalities:

$$g(\hat{a}_j) = E_j, \quad g(a) = \sum_{i=1}^{n} \int_0^{a} \lambda_i(t, \hat{\theta}) Y_i(u) du.$$

Denote by $X_{(1)} \leq ... \leq X_{(n)}$ the ordered sample from $X_1, ..., X_n$. Note that the function

$$g(a) = \sum_{i=1}^{n} \Lambda_i(X_i \wedge a, \hat{\theta})$$

$$= \sum_{i=1}^{n} \left[\sum_{l=i}^{n} \Lambda_{(l)}(a, \hat{\theta}) + \sum_{l=1}^{i-1} \Lambda_{(l)}(X_{(l)}, \hat{\theta}) \right] \mathbf{1}_{[X_{(i-1)}, X_{(i)}]}(a)$$

is continuous and increasing on $[0,\tau]$; here $X_{(0)} = 0$, and we understand $\sum_{l=1}^{0} c_l = 0$. Set

$$b_i = \sum_{l=i+1}^{n} \Lambda_{(l)}(X_{(i)},\hat{\theta}) + \sum_{l=1}^{i} \Lambda_{(l)}(X_{(l)},\hat{\theta}). \qquad [4.20]$$

If $E_j \in [b_{i-1}, b_i]$, then $\hat{a}_j$ is the unique solution of the equation:

$$\sum_{l=i}^{n} \Lambda_{(l)}(\hat{a}_j,\hat{\theta}) + \sum_{l=1}^{i-1} \Lambda_{(l)}(X_{(l)},\hat{\theta}) = E_j. \qquad [4.21]$$

We have $0 < \hat{a}_1 < \hat{a}_2 ... < \hat{a}_k = \tau$.

Under this choice of the intervals $e_j = E_k/k$ for any j.

THEOREM 4.3.– *Under conditions A and B and random choice of the endpoints of grouping intervals, the limit distribution of the statistic Y^2 is chi-square with r degrees of freedom.*

PROOF.– Taking into account that:

$$\frac{E_j}{n} \xrightarrow{P} \frac{j}{k} A(\tau),$$

we have that $\hat{a}_j \xrightarrow{P} a_j$, where a_j is defined by:

$$A(a_j) = \frac{j}{k} A(\tau).$$

By random change of time theorem (see [BIL 68]):

$$(H_n(\hat{a}_1), ..., H_n(\hat{a}_k))^T \xrightarrow{d} (V(a_1), ..., V(a_k))^T,$$

so replacing a_j by $\hat{a}_j$ in the expression of the statistic Y^2, the limit distribution of the statistic Y^2 is chi-square with r degrees of freedom, as in the case of fixed a_j.

The proof is complete.

So, the hypothesis H_0 is rejected with the approximate significance level α if $Y^2 > \chi^2_\alpha(r)$, the statistic Y^2 is computed using formulas [4.19]–[4.21], replacing a_j by $\hat{a}_j$ in all formulas and taking

$$e_1 = ... = e_k = \frac{1}{k}\sum_{i=1}^{n}\Lambda(X_i, \hat{\theta}, z_i).$$

4.4.1. *Test for the exponential AFT model*

Let us consider the exponential AFT model:

$$H_0: \; S_i(t) = \exp\left\{-\int_0^t e^{-\beta^T z_i(u)}\}du\right\}, \; i = 1, ..., n;$$

here $\beta = (\beta_0, \cdots, \beta_m)^T$, $z(u) = (z_0(u), ..., z_m(u))^T$, $z_0(u) \equiv 1$. If z_i are constant, then $H_0: \; S_i(t) = \exp\{-e^{-\beta^T z_i} t\}, \; i = 1, ..., n$.

So $\theta = \beta$ and we have not the nuisance parameter γ in this example ($q = 0$). Note that for exponential model, this is usually the case. This issue was touched on for instance in [PEN 98a].

The maximum likelihood estimator $\hat{\beta}$ of the parameter β maximizes the log-likelihood function:

$$\ell(\beta) = -\sum_{i=1}^{n}\left(\delta_i\beta^T z_i(X_i) + \int_0^{X_i} e^{-\beta^T z_i(u)}du\right).$$

If covariates are constant over time, then

$$\ell(\beta) = -\sum_{i=1}^{n}(\delta_i\beta^T z_i + e^{-\beta^T z_i} X_i).$$

The limits $\hat{a}_j$ are found using formulas [4.19]–[4.21] with:

$$\Lambda_i(t, \hat{\beta}) = \int_0^t e^{-\beta^T z_i(u)}du.$$

If covariates are constant in time, then formulas are explicit; set:

$$b_i = X_{(i)} \sum_{l=i+1}^{n} e^{-\hat{\beta}^T z_{(l)}} + \sum_{l=1}^{i} e^{-\hat{\beta}^T z_{(l)}} X_{(l)}, \quad E_j = \frac{j}{k} \sum_{i=1}^{n} e^{-\hat{\beta}^T z_i} X_i,$$

and denote by i_j the minimal natural number verifying the inequalities $E_j \in [b_{i_j-1}, b_i]$, then

$$\hat{a}_j = \frac{E_j - \sum_{l=1}^{i-1} e^{-\hat{\beta}^T z_{(l)}} X_{(l)}}{\sum_{l=i}^{n} e^{-\hat{\beta}^T z_{(l)}}}, \quad j = 1, ..., k-1, \quad \hat{a}_k = X_{(n)}.$$

By [4.17], the quantities $\hat{i}_{ls}$ and $\hat{C}_{0j}$ (they are used for test statistic computing) are:

$$\hat{i}_{ls} = \frac{1}{n} \sum_{i=1}^{n} \delta_i z_{il}(X_i) z_{is}(X_i),$$

$$\hat{C}_{lj} = -\frac{1}{n} \sum_{i:X_i \in I_j} \delta_i z_{il}(X_i), \quad l, l' = 0, 1, ..., m.$$

Note that the first line of the matrix $\hat{G}$ (see [4.16]) is null: $\hat{g}_{0l'} = 0$, $l' = 0, 1, ..., m$. So, the matrix $\hat{G}$ (and $\hat{G}$) has the rank m and the matrix Σ has the rank $k-1$.

Set $\hat{G}_0 = [\hat{g}_{ll'}]_{m \times m}$, $l, l' = 1, ..., m$, $\tilde{W} = (W_1, ..., W_m)^T$, $\hat{g}_{ll'}$ and W_l are given by [4.16] and [4.18], respectively.

The test statistic is [4.15], where

$$Q = \tilde{W}^T \hat{G}_0^{-1} \tilde{W}, \quad e_j = \frac{1}{k} \sum_{i=1}^{n} \int_0^{X_i} e^{-\beta^T z_i(u)} du.$$

The limit distribution of the statistic is chi-square with $k-1$ degrees of freedom.

The hypothesis H_0 is rejected with the approximate significance level α if $Y_n^2 > \chi^2_\alpha(k-1)$.

4.4.2. *Tests for the scale-shape AFT models with constant covariates*

Let us consider the scale-shape AFT model of the form:

$$S(t|z_i) = S_0\left(\left(e^{-\beta^T z_i} t\right)^{1/\sigma}\right),$$

where S_0 is a specified function, $\sigma > 0$, $\beta = (\beta_0, \beta_1, ..., \beta_m)^T$, $z_i = (z_{i0}, z_{i1}, ..., z_{im})^T$, $z_{i0} = 1$.

Set

$$S(u) = S_0(e^u), \quad h(u) = -\frac{S'(u)}{S(u)},$$

$$v(u) = (\ln h(u))', \quad V_i = \frac{\ln X_i - \beta^T z_i}{\sigma}.$$

In the case of Weibull, loglogistic and lognormal AFT models:

$$\begin{aligned} S_0(t) &= e^{-t}, \quad (1+t)^{-1}, \quad 1-\Phi(\ln t), \\ \ln S(t) &= -e^{-t}, \quad -\ln(1+e^t), \quad \ln(1-\Phi(t)), \\ \ln h(t) &= t, \quad t-\ln(1+e^t), \quad -t^2/2 - \ln\sqrt{2\pi} - \ln(1-\Phi(t)), \\ v(t) &= 1, \quad (1+e^t)^{-1}, \quad -t+\varphi(t)/(1-\Phi(t)), \end{aligned} \qquad [4.22]$$

respectively, where Φ is the cumulative distribution function of the standard normal law. The log-likelihood function is:

$$\ell(\beta, \sigma) = -\delta \ln \sigma + \sum_{i=1}^{n} (\delta_i \ln h(V_i) + \ln S(V_i)).$$

Denote by $\hat{\beta}$ and $\hat{\sigma}$ the ML estimators. Set

$$\hat{V}_i = \frac{\ln X_i - \hat{\beta}^T z_i}{\hat{\sigma}}.$$

The limits $\hat{a}_j$ are found using formulas [4.19]–[4.21] with:

$$\Lambda_i(t, \hat{\beta}) = -\ln S((\ln t - \hat{\beta}^T z_i)/\hat{\sigma}).$$

In the case of the Weibull distribution, the formulas are explicit; set:

$$b_i = X_{(i)}^{1/\hat{\sigma}} \sum_{l=i+1}^{n} e^{-\frac{\hat{\beta}^T z_{(l)}}{\hat{\sigma}}} + \sum_{l=1}^{i} e^{\hat{V}_{(l)}}.$$

If i is the minimal natural number verifying $E_j \in [b_{i-1}, b_i]$, then

$$\hat{a}_j = \left(\frac{E_j - \sum_{l=1}^{i-1} e^{\hat{V}_{(l)}}}{\sum_{l=i}^{n} e^{-\frac{\hat{\beta}^T z_{(l)}}{\hat{\sigma}}}} \right)^{\hat{\sigma}}, \quad j = 1, ..., k-1, \quad \hat{a}_k = X_{(n)}.$$

The expressions for the elements of the matrix $\hat{\mathbf{i}} = [\hat{i}_{ls}]_{(m+2)\times(m+2)}$ are:

$$\hat{i}_{ls} = \frac{1}{n\hat{\sigma}^2} \sum_{i=1}^{n} \delta_i z_{il} z_{is} v^2(\hat{V}_i),$$

$$\hat{i}_{l,m+1} = \frac{1}{n\hat{\sigma}^2} \sum_{i=1}^{n} \delta_i z_{il} v(\hat{V}_i)[1 + \hat{V}_i\, v(\hat{V}_i)], \quad l, s = 0, ..., m.$$

$$\hat{i}_{m+1,m+1} = \frac{1}{n\hat{\sigma}} \sum_{i=1}^{n} \delta_i [1 + \hat{V}_i\, v(\hat{V}_i)]^2, \qquad [4.23]$$

and

$$\hat{C}_{lj} = -\frac{1}{n\hat{\sigma}} \sum_{i:X_i \in I_j} \delta_i z_{il} v(\hat{V}_i), \quad l = 0, ..., m,$$

$$\hat{C}_{m+1,j} = -\frac{1}{n\hat{\sigma}} \sum_{i:X_i \in I_j} \delta_i [1 + \hat{V}_i\, v(\hat{V}_i)]. \qquad [4.24]$$

The estimator $\hat{G}$ has the form:

$$\hat{G} = [\hat{g}_{ll'}], \; l, l' = 0, ..., m+1, \quad \hat{g}_{ll'} = \hat{i}_{ll'} - \sum_{j=1}^{k} \hat{C}_{lj} \hat{C}_{l'j} \hat{A}_j^{-1}. \quad [4.25]$$

Note that in the case of Weibull distribution, the first line of the matrix $\hat{G}$ is null: $\hat{g}_{0l'} = 0$, $l' = 0, 1, ..., m+1$. So, in this case, the matrix $\hat{G}$ (and G) has the rank $m+1$ and the matrix Σ has the rank $k-1$.

In the case of loglogistic and lognormal distributions, the matrix Σ has the rank k, so test statistic is [4.15] with

$$e_j = -\frac{1}{k} \sum_{i=1}^{n} \ln S(\hat{V}_i), \quad \hat{G}^- = \hat{G}^{-1},$$

where $\hat{i}_{ll'}$ and $\hat{C}_{lj}$ are given by formulas [4.23] and [4.24], where v and $\ln S$ are given in [4.22] using the second and the third expressions, respectively, for the loglogistic and lognormal models. Each hypothesis is rejected with the approximate significance level α if $Y^2 > \chi^2_\alpha(k)$.

In the case of Weibull distribution, we define:

$$\hat{G}_0 = [\hat{g}_{ll'}]_{(m+1)\times(m+1)}, \; l, l' = 1, ..., m+1, \quad \tilde{W} = (W_1, ..., W_{m+1})^T. \quad [4.26]$$

The test statistic is [4.15], where

$$e_j = \frac{1}{k} \sum_{i=1}^{n} e^{-\hat{V}_i}, \quad Q = \tilde{W}^T \hat{G}_0^{-1} \tilde{W}.$$

The limit distribution of the statistic is chi-square with $k-1$ degrees of freedom.

EXAMPLE 4.1.– Nelson [NEL 90] gives failure times of 76 electric insulating fluids tested at voltage ranging from 26 to 38 kilovolts (kV).

This experiment was run long enough to observe the failures of all items. The voltage levels v_i, the number of items tested under the same voltage level and the failure times are given in Appendix B, Table B.1.

Diagnostic methods suggest the Weibull AFT-power rule model. Lawless (see [LAW 03]) used parametric methods based on log-likelihood ratios to verify separately Weibull, AFT and power rule assumptions of the model. All tests did not contradict the Weibull AFT-power rule model. Let us first apply the chi-squared test for exponential AFT-power rule model. The ML estimators are $\hat{\beta}_0 = 64.9114$ and $\hat{\beta}_1 = -17.7039$.

Choose $k = 8$ intervals, taking $\tau = 5000$, so we suppose that the data would be censored at this time. The intermediate results given in the below table show that the lengths of the random cells $(\hat{a}_j, \hat{a}_j]$ are very different.

$\hat{a}_j$	1.179	3.729	9.418	18.010	29.746	58.662	131.566	2323.7
U_j	17	13	12	6	5	7	7	9
e_j	9.5	9.5	9.5	9.5	9.5	9.5	9.5	9.5
C_{1j}	-0.797	-0.611	-0.552	-0.275	-0.227	-0.322	-0.315	-0.398

We have $\tilde{W} = -0.1516$, $\hat{i} = 12.2307$. There is only one covariate, so the matrix G_0 has dimension 1. We have:

$$\hat{G}_0 = 0.00446, \ X^2 = \sum_{j=1}^{k} \frac{(U_j - e_j)^2}{U_j} = 12.677, \quad Q = 5.1555,$$

$$Y^2 = X^2 + Q = 17.832, \quad pv_a = P\{\chi^2_7 > 17.832\} = 0.0128.$$

The hypothesis is rejected.

EXAMPLE 4.2.– In Table A2.2 (Appendix 2), survival data of 40 advanced lung cancer patients taken from [LAW 03] are presented. All the patients represented here received prior therapy and were then randomly assigned to one of the two treatments, termed "standard" and "test". Also given for each patient is a Karnofsky score, or performance status, assigned at the time of diagnosis. This is a measure of general medical status on a scale of 10 to 90: 10, 20 and 30 indicates that the patient is completely hospitalized; 40, 50 and 60 indicates that the patient is partially confined to hospital; 70, 80 and 90 indicates that the patient is able to care for himself. So, we use two covariates: treatment and performance status. Starred quantities denote censored observations. Other covariates presented in the initial data are shown (see [LAW 03]) to be not significant. Can the exponential log-linear model (i.e. logarithm of PS considered as the second covariate) be applied to the data?

Choose $k = 6$ grouping intervals. The ML estimators are $\hat{\beta}_0 = 1.1094$, $\hat{\beta}_1 = 0.0596$ and $\hat{\beta}_2 = 0.2683$. The intermediate results are given in the table below.

$\hat{a}_j$	10.01	30.73	65.25	119.34	217.99	999.00
U_j	5	10	5	6	6	8
e_j	6.1667	6.1667	6.1667	6.1667	6.1667	6.1667

We have:

$$\hat{G}_0^{-1} = \begin{pmatrix} 0.002729 & -0.01912 \\ -0.01912 & 4.3593 \end{pmatrix},$$

$$W_0 = (-21.053, -0.1853)^T, \quad Q = 1.20998,$$

$$X^2 = 2.44329, \quad Y^2 = 3.65327, \quad pv = 0.60033.$$

The hypothesis is not rejected.

EXAMPLE 4.3.– Using the data of Example 1, we shall apply the chi-squared test for the Weibull AFT-power rule model. For this model, we take $\beta^T z_i = \beta_0 + \beta_1 z_{i1}$, where $z_{i1} = \ln v_i$. The values of the ML estimators are $\hat{\beta}_0 = 64.85$, $\hat{\beta}_1 = -17.73$ and $\hat{\sigma} = 1.288$.

Choose $k = 8$ intervals. The intermediate results are given in the table below.

$\hat{a}_j$	0.56	1.81	4.63	10.85	21.47	42.57	102.77	2323.7
U_j	7	13	13	10	7	7	7	8
e_j	9.49	9.49	9.49	9.49	9.49	9.49	9.49	9.49
C_{1j}	-0.2550	-0.4741	-0.4724	-0.3563	-0.2477	-0.2509	-0.2812	-0.3772
C_{2j}	0.0915	0.0361	-0.0583	-0.0121	-0.0345	-0.1127	-0.1044	-0.1475

We have:

$$\tilde{W} = (W_1, W_2)^T = (-0.0190, 0.1112)^T,$$

$$\hat{G}_0^{-1} = \begin{pmatrix} 2298.9457 & -152.8077 \\ -152.8077 & 12.337741 \end{pmatrix},$$

$$X^2 = \sum_{j=1}^{k} \frac{(U_j - e_j)^2}{U_j} = 5.0740, \quad Q = \tilde{W}^T \hat{G}_0^{-1} W = 1.6297,$$

$$Y^2 = X^2 + Q = 6.7037, \quad pv_a = P\{\chi_7^2 > 6.7037\} = 0.4604.$$

The data do not contradict the hypothesis.

EXAMPLE 4.4.– Let us consider the lifetimes of 586 patients with the scalar covariate z. All the patients entering the study were older than 69 years and were operated on for small-cell carcinoma of the lung. In Appendix 2, survival times, the censoring indicator and the type of tumor by the TNM classification (Tumor, Nodus and Metastasis) are presented. We shall apply the chi-squared test for the Lognormal PH model.

Using the maximum likelihood method, we construct the proportional hazards model with the cumulative hazard function $\Lambda_x(t; \hat{\beta}, \hat{\theta},) = \Lambda_0(t; \hat{\theta}) \cdot \exp(x\beta)$, where $\Lambda_0(t; \hat{\theta})$ corresponds to the lognormal distribution with the probability density function $f(t, \theta_1, \theta_2) = \frac{1}{\theta_2 t \sqrt{2\pi}} \exp\left(-\frac{1}{2\theta_2^2} \ln^2\left(\frac{t}{\theta_1}\right)\right)$.

The ML estimators of parameters are $\hat{\theta}_1 = 71.55$, $\hat{\theta}_2 = 1.64$ and the regression parameter $\hat{\beta} = 0.26$. The number of estimated parameters of the model is $s = 3$.

To calculate the test statistic Y_n^2, we have chosen the number of interval $k = 7$. The boundary points a_j, E_j, the observed numbers of failures U_j and the expected numbers of failures e_j in the intervals I_j, $j = 1, \ldots, 7$ are presented in the table below.

j	E_j	a_j	U_j	e_j	$\frac{(U_j-e_j)^2}{U_j}$
1	60.64	5.50	58	60.62	0.118
2	121.28	10.57	65	60.64	0.292
3	181.92	17.02	61	60.65	0.002
4	242.55	25.78	60	60.63	0.007
5	303.19	38.79	68	60.64	0.797
6	363.83	61.84	58	60.64	0.120
7	424.47	324.0	54	60.65	0.819

The following matrices were obtained for calculation of Q:

$$Z = \begin{bmatrix} -0.108 \\ 0.180 \\ 0.014 \\ -0.026 \\ 0.304 \\ -0.109 \\ -0.275 \end{bmatrix}, \quad \hat{A} = \begin{bmatrix} 0.099 & 0 & 0 & 0 & 0 & 0 & 0 \\ 0 & 0.111 & 0 & 0 & 0 & 0 & 0 \\ 0 & 0 & 0.104 & 0 & 0 & 0 & 0 \\ 0 & 0 & 0 & 0.102 & 0 & 0 & 0 \\ 0 & 0 & 0 & 0 & 0.116 & 0 & 0 \\ 0 & 0 & 0 & 0 & 0 & 0.099 & 0 \\ 0 & 0 & 0 & 0 & 0 & 0 & 0.092 \end{bmatrix},$$

$$\hat{C} = \begin{bmatrix} -0.002 & -0.001 & -0.001 & -0.001 & -0.001 & -0.001 & -0.001 \\ 0.199 & 0.069 & 0.019 & -0.009 & -0.036 & -0.047 & -0.061 \\ 0.244 & 0.268 & 0.238 & 0.242 & 0.266 & 0.209 & 0.193 \end{bmatrix},$$

$$\hat{G} = \begin{bmatrix} 0.8E-06 & -0.2E-03 & 0.4E-04 \\ -0.2E-03 & 0.07449 & -0.01356 \\ 0.4E-04 & -0.01356 & 0.42602 \end{bmatrix}, \quad W = \begin{bmatrix} -0.0003 \\ 0.0394 \\ 0.0312 \end{bmatrix}.$$

The values of estimates $\hat{i}_{11} = 0.9E-04, \hat{i}_{12} = -0.0037, \hat{i}_{13} = -0.0178, \hat{i}_{22} = -0.5989, \hat{i}_{23} = 0.3601, \hat{i}_{33} = 4.2559$. Thus, $Q = W^T \cdot G^- \cdot W = 1.995$, $\sum_{j=1}^{7} \frac{(U_j-e_j)^2}{U_j} = 2.155$ and $Y_n^2 = 4.15$. The limiting distribution is χ_7^2. The p-value is equal to 0.76, so we have no reason to reject the hypothesis H_0.

4.4.3. *Test for the Weibull AFT model with step-stresses*

In accelerated life testing, one-dimensional step-stresses are often used. Then, the covariates have the form $z_i(u) = (z_{i0}(u), z_{i1}(u))^T$, where $z_{i0}(u) \equiv 1$, $z_{i1}(u)$ are step-stresses of the form:

$$z_{i1}(u) = \begin{cases} z_{i11}, & 0 \le u < t_{i1}, \\ z_{i12}, & t_{i1} \le u < t_{i2}. \\ \cdots & \cdots \\ z_{i1\nu}, & t_{i,\nu-1} \le u < t_{i\nu_i}; \end{cases}$$

here z_{i1j} are constant one-dimensional stresses.

Let us consider goodness-of-fit for the Weibull AFT model with step-stresses:

$$S(t|z_i) = e^{-(\int_0^t e^{-\beta^T z_i(u)} du)^{1/\sigma}}.$$

The log-likelihood function is:

$$\ell(\beta, \sigma) = \sum_{i=1}^n \delta_i \{-\ln \sigma - \beta^T z_i(X_i) + (1/\sigma - 1)$$

$$\ln \int_0^{X_i} e^{-\beta^T z_i(u)} du\} - \sum_{i=1}^n (\int_0^{X_i} e^{-\beta^T z_i(u)} du)^{1/\sigma}.$$

Set

$$\hat{V}_i = \frac{1}{\hat{\sigma}} \ln(\int_0^{X_i} e^{-\hat{\beta}^T z_i(u)} du),$$

$$z_l^{(i)} = \frac{\int_0^{X_i} z_{il}(u) e^{-\hat{\beta}^T z_i(u)} du}{\int_0^{X_i} e^{-\hat{\beta}^T z_i(u)} du}, \qquad l = 0, 1.$$

Note that $z_0^{(i)} = z_{i0}(u) \equiv 1$ and all integrals are sums; if $t \in [t_{i,j-1}, t_{ij})$, $j = 1, \cdots, \nu_i$, then

$$\int_0^t e^{-\hat{\beta}^T z_i(u)} du = \mathbf{1}_{\{j>1\}} \sum_{s=1}^{j-1} e^{-\hat{\beta}^T z_{is}} (t_{is} - t_{i,s-1}) + e^{-\beta^T z_{ij}} (t - t_{i,j-1}),$$

and

$$\int_0^t z_{il}(u)e^{-\hat{\beta}^T z_i(u)}du$$

$$= \mathbf{1}_{\{j>1\}} \sum_{s=1}^{j-1} z_{i1s}\, e^{-\hat{\beta}^T z_{is}}(t_{is} - t_{i,s-1}) + e^{-\hat{\beta}^T z_{ij}}(t - t_{i,j-1}).$$

The limits $\hat{a}_j$ are found using formulas [4.19]–[4.21] with

$$\Lambda_i(t, \hat{\beta}) = (\int_0^t e^{-\hat{\beta}^T z_i(u)}du)^{1/\hat{\sigma}}.$$

The expressions for the elements of the matrix $\hat{\mathbf{i}} = [\hat{i}_{ls}]_{3\times 3}$ are:

$$\hat{i}_{ls} = \frac{1}{n\hat{\sigma}^2}\sum_{i=1}^n \delta_i[z_l^{(i)} + \hat{\sigma}(z_{il}(X_i) - z_l^{(i)})][z_s^{(i)} + \hat{\sigma}(z_{is}(X_i) - z_s^{(i)})],$$

$$\hat{i}_{l2} = \frac{1}{n\hat{\sigma}^2}\sum_{i=1}^n \delta_i[z_l^{(i)} + \hat{\sigma}(z_{il}(X_i) - z_l^{(i)})][1 + \hat{V}_i], \quad l, s = 0, 1.$$

$$\hat{i}_{22} = \frac{1}{n\hat{\sigma}}\sum_{i=1}^n \delta_i[1 + \hat{V}_i]^2,$$

and

$$\hat{C}_{lj} = -\frac{1}{n\hat{\sigma}}\sum_{i:X_i \in I_j} \delta_i[z_l^{(i)} + \hat{\sigma}(z_{il}(X_i) - z_l^{(i)})], \quad l = 0, 1,$$

$$\hat{C}_{2j} = -\frac{1}{n\hat{\sigma}}\sum_{i:X_i \in I_j} \delta_i[1 + \hat{V}_i].$$

The estimator $\hat{G}$ has the form [4.25] and the first line of the matrix $\hat{G}$ is null: $\hat{g}_{0l'} = 0$, $l' = 0, 1, 2$. So, in this case, the matrix $\hat{G}$ (and G) has rank 2 and the matrix Σ has the rank $k - 1$.

The test statistic is [4.15], where

$$e_j = \frac{1}{k}\sum_{i=1}^{n}\left(\int_0^{X_i} e^{-\hat{\beta}^T z_i(u)}du\right)^{1/\hat{\sigma}}, \quad Q = \tilde{W}^T \hat{G}_0^{-1} \tilde{W},$$

and $\hat{G}_0$, $\tilde{W}$ have the form [4.26] with $m = 1$. The limit distribution of the statistic is chi-square with $k - 1$ degrees of freedom.

APPENDICES

Appendix 1

Tables of Asymptotically D-optimal Grouping

k	x_1	x_2	x_3	x_4	x_5	x_6	x_7	x_8	x_9	x_{10}	A
2	1.5936	–	–	–	–	–	–	–	–	–	0.6476
3	1.0176	2.6112	–	–	–	–	–	–	–	–	0.8203
4	0.7541	1.7716	3.3652	–	–	–	–	–	–	–	0.891
5	0.6004	1.3545	2.372	3.9657	–	–	–	–	–	–	0.9269
6	0.4993	1.0997	1.8538	2.8714	4.465	–	–	–	–	–	0.9606
7	0.4276	0.9269	1.5273	2.2813	3.2989	4.8925	–	–	–	–	0.9606
8	0.3739	0.8015	1.3008	1.9012	2.6553	3.6729	5.2665	–	–	–	0.9693
9	0.3323	0.7063	1.1338	1.6331	2.2336	2.9876	4.0052	5.5988	–	–	0.9754
10	0.299	0.6314	1.0053	1.4329	1.9322	2.5326	3.2866	4.3042	5.8979	–	0.9798
11	0.2716	0.5695	0.9014	1.2746	1.7015	2.1989	2.7955	3.5429	4.548	6.1176	0.9832

Table A1.1. *D-optimal boundary points of grouping intervals for the exponential distribution and corresponding values of the relative asymptotic information A*

k	p_1	p_2	p_3	p_4	p_5	p_6	p_7	p_8	p_9	p_{10}	p_{11}
2	0.7968	0.2032	–	–	–	–	–	–	–	–	–
3	0.6385	0.288	0.0735	–	–	–	–	–	–	–	–
4	0.5296	0.3004	0.1355	0.0345	–	–	–	–	–	–	–
5	0.4514	0.2905	0.1648	0.0744	0.0189	–	–	–	–	–	–
6	0.393	0.274	0.1763	0.1	0.0451	0.0116	–	–	–	–	–
7	0.3479	0.2563	0.1787	0.115	0.0652	0.0294	0.0075	–	–	–	–
8	0.312	0.2394	0.1763	0.1229	0.0791	0.0449	0.0202	0.0052	–	–	–
9	0.2827	0.2238	0.1717	0.1265	0.0882	0.0567	0.0322	0.0145	0.0037	–	–
10	0.2584	0.2097	0.1659	0.1273	0.0938	0.0654	0.0421	0.0239	0.0107	0.0028	–
11	0.2378	0.1964	0.1598	0.1264	0.0971	0.0715	0.0498	0.0322	0.0183	0.0083	0.0024

Table A1.2. *D-optimal probabilities $P\{X_1 \in I_j\}, j = 1, ..., k$ for the exponential and Rayleigh distributions*

k	x_1	x_2	x_3	x_4	x_5	x_6	x_7	x_8	x_9	A
2	1.7853	–	–	–	–	–	–	–	–	0.6476
3	1.4266	2.2853	–	–	–	–	–	–	–	0.8203
4	1.228	1.8823	2.5943	–	–	–	–	–	–	0.891
5	1.0954	1.6459	2.1781	2.8163	–	–	–	–	–	0.9269
6	0.9993	1.4831	1.9255	2.3964	2.9883	–	–	–	–	0.9476
7	0.9247	1.3615	1.7477	2.136	2.5686	3.1281	–	–	–	0.9606
8	0.8648	1.2661	1.613	1.95	2.3045	2.7103	3.2455	–	–	0.9693
9	0.8152	1.1884	1.5058	1.8073	2.1136	2.4444	2.8303	3.3463	–	0.9754
10	0.7734	1.1237	1.418	1.6928	1.9658	2.2506	2.5638	2.934	3.4344	0.9798

Table A1.3. *D-optimal boundary points of grouping intervals for the Rayleigh distribution and corresponding values of the relative asymptotic information A*

k	x_1	x_2	x_3	x_4	x_5	x_6	x_7	x_8	x_9	A
2	2.0451	–	–	–	–	–	–	–	–	0.6451
3	1.6762	2.5366	–	–	–	–	–	–	–	0.8179
4	1.4689	2.1292	2.8402	–	–	–	–	–	–	0.8892
5	1.3292	1.8879	2.4221	3.0583	–	–	–	–	–	0.9254
6	1.2261	1.7205	2.1667	2.6379	3.2274	–	–	–	–	0.9464
7	1.1458	1.5947	1.9859	2.3759	2.8081	3.3649	–	–	–	0.9596
8	1.0807	1.4952	1.8481	2.1879	2.5431	2.948	3.4803	–	–	0.9685
9	1.0267	1.4138	1.7377	2.0423	2.3499	2.6803	3.0652	3.5789	–	0.9747
10	0.9798	1.3447	1.646	1.9252	2.2003	2.4857	2.7984	3.1668	3.6625	0.9792

Table A1.4. *D-optimal boundary points of grouping intervals for the Maxwell distribution and corresponding values of the relative asymptotic information A*

k	p_1	p_2	p_3	p_4	p_5	p_6	p_7	p_8	p_9	p_{10}
2	0.7576	0.2424	–	–	–	–	–	–	–	–
3	0.5781	0.3296	0.0923	–	–	–	–	–	–	–
4	0.4597	0.331	0.1647	0.0446	–	–	–	–	–	–
5	0.3778	0.3097	0.1942	0.0933	0.025	–	–	–	–	–
6	0.3185	0.2837	0.2022	0.1223	0.0579	0.0154	–	–	–	–
7	0.2739	0.2585	0.2	0.1374	0.0818	0.0383	0.0101	–	–	–
8	0.2393	0.2357	0.1931	0.1438	0.0971	0.0573	0.0267	0.007	–	–
9	0.2119	0.2155	0.184	0.145	0.1063	0.0711	0.0418	0.0194	0.005	–
10	0.1891	0.1978	0.1745	0.1437	0.1112	0.0806	0.0536	0.0313	0.0145	0.0038

Table A1.5. *D-optimal probabilities $P\{X_1 \in I_j\}, j = 1, ..., k$ for the Maxwell distribution*

k	x_1	x_2	x_3	x_4	x_5	x_6	x_7	x_8	x_9	A
2	1.4821	–	–	–	–	–	–	–	–	0.6523
3	1.1395	1.9952	–	–	–	–	–	–	–	0.8244
4	0.9553	1.5993	2.3124	–	–	–	–	–	–	0.8943
5	0.8355	1.3712	1.9002	2.5407	–	–	–	–	–	0.9295
6	0.7495	1.2163	1.6526	2.1225	2.7173	–	–	–	–	0.9496
7	0.6842	1.1025	1.4797	1.8658	2.3009	2.8632	–	–	–	0.9622
8	0.6277	1.0069	1.3432	1.6751	2.0278	2.4343	2.9742	–	–	0.9706
9	0.5892	0.9409	1.2488	1.5445	1.8503	2.1861	2.5844	3.1015	–	0.9765
10	0.5545	0.8838	1.1682	1.438	1.7113	2.0012	2.3249	2.6993	3.2125	0.9808

Table A1.6. *D-optimal boundary points of grouping intervals for the half-normal distribution and corresponding values of the relative asymptotic information* A

k	p_1	p_2	p_3	p_4	p_5	p_6	p_7	p_8	p_9	p_{10}
2	0.8617	0.1383	–	–	–	–	–	–	–	–
3	0.7455	0.2085	0.046	–	–	–	–	–	–	–
4	0.6606	0.2297	0.089	0.0207	–	–	–	–	–	–
5	0.5966	0.2331	0.1129	0.0463	0.0111	–	–	–	–	–
6	0.5464	0.2297	0.1255	0.0646	0.0272	0.0066	–	–	–	–
7	0.5061	0.2236	0.1313	0.0769	0.0407	0.0172	0.0042	–	–	–
8	0.4698	0.2162	0.1348	0.0853	0.0513	0.0277	0.012	0.0029	–	–
9	0.4443	0.209	0.135	0.0893	0.0582	0.0355	0.019	0.0078	0.0019	–
10	0.4208	0.2024	0.1341	0.0923	0.0634	0.0417	0.0253	0.0131	0.0056	0.0013

Table A1.7. *D-optimal probabilities* $P\{X_1 \in I_j\}, j = 1, ..., k$ *for the half-normal distribution*

k	x_1	x_2	x_3	x_4	x_5	x_6	x_7	x_8	x_9	x_{10}	x_{11}	x_{12}	x_{13}	x_{14}	A
3	0.2731	2.6067	–	–	–	–	–	–	–	–	–	–	–	–	0.4079
4	0.2109	1.3979	3.4137	–	–	–	–	–	–	–	–	–	–	–	0.5572
5	0.1044	0.5123	1.959	3.8606	–	–	–	–	–	–	–	–	–	–	0.6836
6	0.0772	0.3649	1.2269	2.5726	4.4096	–	–	–	–	–	–	–	–	–	0.7571
7	0.0501	0.2318	0.6758	1.7192	2.9922	4.7959	–	–	–	–	–	–	–	–	0.8109
8	0.0377	0.174	0.4837	1.1904	2.2041	3.4285	5.2049	–	–	–	–	–	–	–	0.848
9	0.0275	0.1269	0.3431	0.7829	1.6027	2.5713	3.7667	5.5273	–	–	–	–	–	–	0.8756
10	0.0213	0.0988	0.2638	0.577	1.1805	1.9932	2.9269	4.1024	5.8478	–	–	–	–	–	0.8963
11	0.0165	0.0771	0.2046	0.4359	0.856	1.5344	2.3192	3.2319	4.393	6.127	–	–	–	–	0.9123
12	0.0123	0.0618	0.1638	0.3434	0.6517	1.1789	1.857	2.6163	3.5103	4.6589	6.3853	–	–	–	0.9248
13	0.0106	0.05	0.1326	0.2754	0.5106	0.903	1.4807	2.1401	2.881	3.7623	4.9016	6.6208	–	–	0.9349
14	0.0087	0.0412	0.1094	0.2261	0.3126	0.7116	1.1798	1.7608	2.4019	3.1286	3.9997	5.1314	6.8444	–	0.9431
15	0.0072	0.0344	0.0913	0.1881	0.3394	0.5734	0.9387	1.4426	2.0116	2.6381	3.3538	4.2169	5.3425	7.0506	0.9498

Table A1.8. *D-optimal boundary points of grouping intervals for the Weibull and minimum extreme value distributions and corresponding values of the relative asymptotic information A*

k	p_1	p_2	p_3	p_4	p_5	p_6	p_7	p_8	p_9	p_{10}	p_{11}	p_{12}	p_{13}	p_{14}	p_{15}
3	0.239	0.6872	0.0738	–	–	–	–	–	–	–	–	–	–	–	–
4	0.1901	0.5628	0.2142	0.0329	–	–	–	–	–	–	–	–	–	–	–
5	0.0991	0.3018	0.4581	0.1199	0.0211	–	–	–	–	–	–	–	–	–	–
6	0.0743	0.2314	0.4011	0.2169	0.0641	0.0122	–	–	–	–	–	–	–	–	–
7	0.0489	0.1581	0.2843	0.3295	0.129	0.0419	0.0083	–	–	–	–	–	–	–	–
8	0.037	0.1227	0.2238	0.3124	0.1938	0.0779	0.0269	0.0055	–	–	–	–	–	–	–
9	0.0271	0.0921	0.1712	0.2525	0.2557	0.125	0.0533	0.0191	0.004	–	–	–	–	–	–
10	0.0211	0.0729	0.1379	0.2065	0.2545	0.1708	0.0827	0.0371	0.0136	0.0029	–	–	–	–	–
11	0.0164	0.0578	0.1108	0.1683	0.2218	0.2101	0.1164	0.0589	0.0271	0.0102	0.0022	–	–	–	–
12	0.0131	0.0468	0.0912	0.1395	0.1882	0.2136	0.1515	0.083	0.0432	0.0204	0.0078	0.0017	–	–	–
13	0.0105	0.0383	0.0754	0.1165	0.1592	0.1947	0.1779	0.1099	0.0615	0.0329	0.0158	0.0061	0.0013	–	–
14	0.0087	0.0317	0.0632	0.0988	0.1357	0.171	0.1836	0.1354	0.0814	0.0467	0.0255	0.0124	0.0048	0.0011	–
15	0.0072	0.0266	0.0535	0.0842	0.1163	0.1486	0.1725	0.1548	0.1025	0.0623	0.0365	0.0203	0.0099	0.0039	0.0009

Table A1.9. *D-optimal probabilities $P\{X_1 \in I_j\}, j = 1, ..., k$ for the Weibull and minimum extreme value distributions*

k	x_1	x_2	x_3	x_4	x_5	x_6	x_7	x_8	x_9	x_{10}	x_{11}	x_{12}	x_{13}	x_{14}	A
3	–0.9581	1.2978	–	–	–	–	–	–	–	–	–	–	–	–	0.4079
4	–1.2278	–0.335	1.5562	–	–	–	–	–	–	–	–	–	–	–	0.5572
5	–1.3508	–0.6725	0.6688	2.2591	–	–	–	–	–	–	–	–	–	–	0.6836
6	–1.4838	–0.9449	–0.2045	1.008	2.5608	–	–	–	–	–	–	–	–	–	0.7571
7	–1.5676	–1.096	–0.5419	0.3919	1.4614	2.9934	–	–	–	–	–	–	–	–	0.8109
8	–1.6496	–1.2321	–0.7903	–0.1743	0.7262	1.749	3.2777	–	–	–	–	–	–	–	0.848
9	–1.7097	–1.3265	–0.9444	–0.4717	0.2448	1.0698	2.0644	3.5952	–	–	–	–	–	–	0.8756
10	–1.766	–1.4115	–1.074	–0.6897	–0.1659	0.5499	1.3326	2.3149	3.8496	–	–	–	–	–	0.8963
11	–1.8119	–1.4787	–1.1715	–0.8397	–0.4262	0.1576	0.5877	1.5877	2.5639	4.104	–	–	–	–	0.9123
12	–1.854	–1.5388	–1.2557	–0.9617	–0.619	–0.1648	0.4282	1.0689	1.8092	2.7833	4.328	–	–	–	0.9248
13	–1.8902	–1.5896	–1.325	–1.0581	–0.7608	–0.3925	0.102	0.6723	1.2894	2.0205	2.9948	4.5454	–	–	0.9349
14	–1.9234	–1.6354	–1.3862	–1.1406	–0.8763	–0.5658	–0.1654	0.3401	0.8852	1.4868	2.2126	3.1883	4.7419	–	0.9431
15	–1.9531	–1.6757	–1.4391	–1.2101	–0.9701	–0.699	–0.3665	0.0632	0.5564	1.0811	1.6723	2.3949	3.3713	4.9279	0.9498

Table A1.10. *D-optimal boundary points of grouping intervals for the maximum extreme value distribution and corresponding values of the relative asymptotic information A*

k	p_1	p_2	p_3	p_4	p_5	p_6	p_7	p_8	p_9	p_{10}	p_{11}	p_{12}	p_{13}	p_{14}	p_{15}
3	0.0738	0.6872	0.239	–	–	–	–	–	–	–	–	–	–	–	–
4	0.0329	0.2142	0.5628	0.1901	–	–	–	–	–	–	–	–	–	–	–
5	0.0211	0.1199	0.4581	0.3018	0.0991	–	–	–	–	–	–	–	–	–	–
6	0.0122	0.0641	0.2169	0.4011	0.2314	0.0743	–	–	–	–	–	–	–	–	–
7	0.0083	0.0419	0.129	0.3295	0.2843	0.1581	0.0489	–	–	–	–	–	–	–	–
8	0.0055	0.0269	0.0779	0.1938	0.3124	0.2238	0.1227	0.037	–	–	–	–	–	–	–
9	0.004	0.0191	0.0533	0.125	0.2557	0.2525	0.1712	0.0921	0.0271	–	–	–	–	–	–
10	0.0029	0.0136	0.0371	0.0827	0.1708	0.2545	0.2065	0.1379	0.0729	0.0211	–	–	–	–	–
11	0.0022	0.0102	0.0271	0.0589	0.1164	0.2101	0.2218	0.1683	0.1108	0.0578	0.0164	–	–	–	–
12	0.0017	0.0078	0.0204	0.0432	0.083	0.1515	0.2136	0.1882	0.1395	0.0912	0.0463	0.0131	–	–	–
13	0.0013	0.0061	0.0158	0.0329	0.0615	0.1099	0.1779	0.1947	0.1592	0.1165	0.0754	0.0383	0.0105	–	–
14	0.0011	0.0048	0.0124	0.0255	0.0467	0.0814	0.1354	0.1836	0.171	0.1357	0.0988	0.0632	0.0317	0.0087	–
15	0.0009	0.0039	0.0099	0.0203	0.0365	0.0623	0.1025	0.1548	0.1725	0.1486	0.1163	0.0842	0.0535	0.0266	0.0072

Table A1.11. *D-optimal probabilities* $P\{X_1 \in I_j\}, j = 1, ..., k$ *for the maximum extreme value distribution*

k	x_1	x_2	x_3	x_4	x_5	x_6	x_7	x_8	x_9	x_{10}	x_{11}	x_{12}	x_{13}	x_{14}	A
3	–1.1106	1.1106	–	–	–	–	–	–	–	–	–	–	–	–	0.4065
4	–1.3834	0	1.3834	–	–	–	–	–	–	–	–	–	–	–	0.5527
5	–1.6961	–0.6894	0.6894	1.6961	–	–	–	–	–	–	–	–	–	–	0.6826
6	–1.8817	–0.997	0	0.997	1.8817	–	–	–	–	–	–	–	–	–	0.7557
7	–2.06	–1.2647	–0.4918	0.4918	1.2647	2.06	–	–	–	–	–	–	–	–	0.8103
8	–2.1954	–1.4552	–0.7863	0	0.7863	1.4552	2.1954	–	–	–	–	–	–	–	0.8474
9	–2.3188	–1.6218	–1.0223	–0.3828	0.3828	1.0223	1.6218	2.3188	–	–	–	–	–	–	0.8753
10	–2.4225	–1.7578	–1.2046	–0.6497	0	0.6497	1.2046	1.7578	2.4225	–	–	–	–	–	0.896
11	–2.5167	–1.8784	–1.3602	–0.8621	–0.3143	0.3143	0.8621	1.3602	1.8784	2.5167	–	–	–	–	0.9121
12	–2.5993	–1.9028	–1.4914	–1.0331	–0.5334	0	0.5334	1.0331	1.4914	1.9028	2.5993	–	–	–	0.9247
13	–2.6746	–2.0762	–1.6068	–1.1784	–0.7465	–0.2669	0.2669	0.7465	1.1784	1.6068	2.0762	2.6746	–	–	0.9348
14	–2.7436	–2.1609	–1.7092	–1.3042	–0.9065	–0.4818	0	0.4818	0.9065	1.3042	1.7092	2.1609	2.7436	–	0.943
15	–2.8069	–2.2378	–1.8011	–1.415	–1.0435	–0.659	–0.2325	0.2325	0.659	1.0435	1.415	1.8011	2.2378	2.8069	0.9498

Table A1.12. *D-optimal boundary points of grouping intervals for the normal and lognormal distributions and corresponding values of the relative asymptotic information A*

k	p_1	p_2	p_3	p_4	p_5	p_6	p_7	p_8	p_9	p_{10}	p_{11}	p_{12}	p_{13}	p_{14}	p_{15}
3	0.1334	0.7332	0.1334	–	–	–	–	–	–	–	–	–	–	–	–
4	0.0833	0.4167	0.4167	0.0833	–	–	–	–	–	–	–	–	–	–	–
5	0.0449	0.2004	0.5094	0.2004	0.0449	–	–	–	–	–	–	–	–	–	–
6	0.0299	0.1295	0.3406	0.3406	0.1295	0.0299	–	–	–	–	–	–	–	–	–
7	0.0197	0.0833	0.2084	0.3772	0.2084	0.0833	0.0197	–	–	–	–	–	–	–	–
8	0.0141	0.0587	0.1431	0.2841	0.2841	0.1431	0.0587	0.0141	–	–	–	–	–	–	–
9	0.0102	0.0422	0.1009	0.1976	0.2982	0.1976	0.1009	0.0422	0.0102	–	–	–	–	–	–
10	0.0077	0.0317	0.0748	0.1438	0.242	0.242	0.1438	0.0748	0.0317	0.0077	–	–	–	–	–
11	0.0059	0.0243	0.0567	0.1074	0.1823	0.2468	0.1823	0.1074	0.0567	0.0243	0.0059	–	–	–	–
12	0.0047	0.019	0.0442	0.0829	0.1392	0.21	0.21	0.1392	0.0829	0.0442	0.019	0.0047	–	–	–
13	0.0037	0.0152	0.0352	0.0652	0.1085	0.167	0.2104	0.167	0.1085	0.0652	0.0352	0.0152	0.0037	–	–
14	0.003	0.0124	0.0283	0.0524	0.0862	0.1327	0.185	0.185	0.1327	0.0862	0.0524	0.0283	0.0124	0.003	-
15	0.0025	0.0101	0.0232	0.0427	0.0698	0.1066	0.1532	0.1838	0.1532	0.1066	0.0698	0.0427	0.0232	0.0101	0.0025

Table A1.13. *D-optimal probabilities* $P\{X_1 \in I_j\}, j = 1, ..., k$ *for the normal and lognormal distributions*

k	x_1	x_2	x_3	x_4	x_5	x_6	x_7	x_8	x_9	x_{10}	x_{11}	x_{12}	x_{13}	x_{14}	A
3	–1.1674	1.1674	–	–	–	–	–	–	–	–	–	–	–	–	0.4426
4	–1.9797	0	1.9797	–	–	–	–	–	–	–	–	–	–	–	0.6026
5	–2.509	–0.847	0.847	2.509	–	–	–	–	–	–	–	–	–	–	0.6861
6	–2.9083	–1.3335	0	1.3335	2.9083	–	–	–	–	–	–	–	–	–	0.789
7	–3.2797	–1.7426	–0.5955	0.5955	1.7426	3.2797	–	–	–	–	–	–	–	–	0.8369
8	–3.6023	–2.0786	–1.0165	0	1.0165	2.0786	3.6023	–	–	–	–	–	–	–	0.8701
9	–3.8981	–2.3777	–1.3604	–0.4631	0.4631	1.3604	2.3777	3.8981	–	–	–	–	–	–	0.8942
10	–4.1665	–2.6442	–1.6512	–0.8235	0	0.8235	1.6512	2.6442	4.1665	–	–	–	–	–	0.9123
11	–4.4148	–2.8883	–1.9085	–1.1227	–0.3805	0.3805	1.1227	1.9085	2.8883	4.4148	–	–	–	–	0.9261
12	–4.6437	–3.1123	–2.1394	–1.3798	–0.6929	0	0.6929	1.3798	2.1394	3.1123	4.6437	–	–	–	0.9371
13	–4.8577	–3.3212	–2.3511	–1.6081	–0.958	–0.3225	0.3235	0.958	1.6081	2.3511	3.3212	4.8577	–	–	0.9457
14	–5.0564	–3.515	–2.546	–1.8138	–1.1888	–0.5984	0	0.5984	1.1888	1.8138	2.546	3.515	5.0564	–	0.9526
15	–5.2452	–3.6987	–2.7289	–2.0032	–1.3953	–0.8364	–0.2818	0.2818	0.8364	1.3953	2.0032	2.4289	3.6987	5.2452	0.9585

Table A1.14. *D-optimal boundary points of grouping intervals for the logistic and loglogistic distributions and corresponding values of the relative asymptotic information* A

k	p_1	p_2	p_3	p_4	p_5	p_6	p_7	p_8	p_9	p_{10}	p_{11}	p_{12}	p_{13}	p_{14}	p_{15}
3	0.2373	0.5254	0.2373	–	–	–	–	–	–	–	–	–	–	–	–
4	0.1214	0.3786	0.3786	0.1214	–	–	–	–	–	–	–	–	–	–	–
5	0.0752	0.2249	0.3998	0.2249	0.0752	–	–	–	–	–	–	–	–	–	–
6	0.0517	0.1568	0.2915	0.2915	0.1568	0.0517	–	–	–	–	–	–	–	–	–
7	0.0363	0.1127	0.2064	0.2892	0.2064	0.1127	0.0363	–	–	–	–	–	–	–	–
8	0.0265	0.0847	0.1545	0.2343	0.2343	0.1545	0.0847	0.0265	–	–	–	–	–	–	–
9	0.0199	0.065	0.1193	0.1821	0.2274	0.1821	0.1193	0.065	0.0199	–	–	–	–	–	–
10	0.0153	0.051	0.0946	0.1441	0.195	0.195	0.1441	0.0946	0.051	0.0153	–	–	–	–	–
11	0.0119	0.0408	0.0764	0.1164	0.1605	0.188	0.1605	0.1164	0.0764	0.0408	0.0119	–	–	–	–
12	0.0095	0.0331	0.0627	0.0957	0.1324	0.1666	0.1666	0.1324	0.0957	0.0627	0.0331	0.0095	–	–	–
13	0.0077	0.0271	0.0521	0.0799	0.1104	0.1425	0.1606	0.1425	0.1104	0.0799	0.0521	0.0271	0.0077	–	–
14	0.0063	0.0226	0.0438	0.0675	0.0933	0.1212	0.1453	0.1453	0.1212	0.0933	0.0675	0.0438	0.0226	0.0063	–
15	0.0053	0.0189	0.0371	0.0576	0.0797	0.1037	0.1277	0.14	0.1277	0.1037	0.0797	0.0576	0.0371	0.0189	0.0053

Table A1.15. *D-optimal probabilities* $P\{X_1 \in I_j\}, j = 1, ..., k$ *for the logistic and loglogistic distributions*

Appendix 2

Data Sets

z_i (kV)	Frequency	Failure times T_i
26	3	5.79 1579.52 2323.7
28	5	68.85 426.07 110.29 108.29 1067.6
30	11	17.05 22.66 21.02 175.88 139.07 144.12 20.46 43.40 194.90 47.30 7.74
32	15	0.40 82.85 9.88 89.29 215.10 2.75 0.79 15.93 3.91 0.27 0.69 100.58 27.80 13.95 53.24
34	19	0.96 4.15 0.19 0.78 8.01 31.75 7.35 6.50 8.27 33.91 32.52 3.16 4.85 2.78 4.67 1.31 12.06 36.71 72.89
36	15	1.97 0.59 2.58 1.69 2.71 25.50 0.35 0.99 3.99 3.67 2.07 0.96 5.35 2.90 13.77
38	8	0.47 0.73 1.40 0.74 0.39 1.13 0.09 2.38

Table A2.1. *The failure times of 76 electric insulating fluids, see [NEL 90]*

PS	Lifetimes X_i, standard treatment	Lifetimes X_i, test treatment
20	8	1
30	16	20 51 19
40	82 8 21 10 12	2 18
50	12 177	231* 15
60	126 153 287 100	44 90 43
70	411 118 25* 11 250	991 103* 164 231
80	54 56 200	201 84 340
90		999

Table A2.2. *Survival data on 40 advanced lung cancer patients, see [LAW 03]*

i	X_i	δ	z_i	i	X_i	δ	z_i	i	X_i	δ	z_i	i	X_i	δ	z_i	i	X_i	δ	z_i
1	1	1	2	119	10	1	3	237	61	1	3	355	5	1	3	473	17	1	2
2	187	0	2	120	117	1	2	238	64	0	1	356	74	0	3	474	11	1	1
3	5	1	3	121	37	1	2	239	11	1	3	357	55	1	2	475	23	1	1
4	122	0	2	122	98	1	2	240	68	1	4	358	69	0	1	476	4	1	1
5	54	0	1	123	14	1	2	241	21	1	4	359	30	1	1	477	26	1	3
6	20	1	3	124	9	1	3	242	33	0	3	360	70	0	3	478	55	1	1
7	2	0	2	125	61	1	2	243	29	1	3	361	69	0	2	479	19	1	2
8	160	0	1	126	60	0	2	244	140	0	3	362	66	0	2	480	10	1	3
9	38	1	3	127	24	1	1	245	21	1	3	363	17	1	2	481	2	0	1
10	62	1	1	128	17	1	2	246	13	1	2	364	41	1	3	482	19	1	3
11	184	0	2	129	37	1	2	247	63	1	3	365	22	1	3	483	1	1	4
12	13	0	1	130	61	0	2	248	12	1	3	366	7	1	3	484	18	1	2
13	26	1	2	131	23	1	2	249	26	1	3	367	29	1	2	485	2	1	2
14	28	0	2	132	2	1	3	250	173	0	2	368	1	0	2	486	86	1	3
15	10	1	1	133	82	1	3	251	13	1	3	369	63	0	2	487	5	1	3
16	5	1	3	134	131	0	2	252	3	1	2	370	63	0	2	488	10	1	3
17	9	1	2	135	19	1	2	253	39	1	3	371	3	1	3	489	99	0	2
18	40	1	2	136	1	1	3	254	19	1	3	372	32	1	2	490	16	1	3
19	5	1	2	137	86	1	3	255	55	1	3	373	22	1	1	491	23	1	3
20	27	0	1	138	14	1	3	256	3	1	3	374	10	1	3	492	4	1	3
21	106	1	2	139	65	0	1	257	49	1	1	375	67	0	3	493	24	1	3
22	79	1	3	140	39	1	2	258	23	1	2	376	4	1	3	494	73	1	3
23	26	1	2	141	18	1	2	259	16	1	1	377	38	1	4	495	29	1	4
24	7	1	2	142	60	0	3	260	44	1	2	378	62	0	3	496	88	0	3
25	6	1	1	143	212	0	3	261	27	1	3	379	61	0	2	497	100	0	3
26	4	1	2	144	14	1	3	262	145	0	3	380	61	0	3	498	73	1	3
27	86	1	2	145	11	1	3	263	5	1	2	381	14	1	2	499	87	1	3
28	137	0	1	146	19	1	3	264	28	1	3	382	16	1	3	500	36	1	3
29	55	0	2	147	48	1	3	265	130	1	1	383	17	1	3	501	29	1	3
30	127	1	2	148	60	1	3	266	8	1	2	384	16	1	1	502	31	0	2
31	10	1	2	149	18	1	3	267	50	1	3	385	2	1	2	503	92	0	1
32	324	0	2	150	7	1	2	268	19	1	3	386	80	1	2	504	8	1	3
33	25	0	1	151	2	1	1	269	70	1	1	387	30	0	3	505	50	1	2
34	174	0	1	152	65	0	1	270	16	1	1	388	59	0	2	506	56	1	3
35	14	1	2	153	5	1	2	271	125	0	3	389	8	1	2	507	90	0	3
36	6	1	3	154	22	1	2	272	11	1	3	390	7	1	3	508	80	1	1
37	48	1	2	155	22	1	4	273	112	0	3	391	60	1	2	509	37	1	2

i	X_i	δ	z_i	i	X_i	δ	z_i	i	X_i	δ	z_i	i	X_i	δ	z_i	i	X_i	δ	z_i
38	41	1	2	156	149	1	3	274	216	0	3	392	54	0	2	510	20	1	2
39	7	1	2	157	5	1	3	275	35	0	2	393	8	1	3	511	8	1	3
40	175	0	1	158	6	1	3	276	31	1	3	394	52	0	3	512	68	0	2
41	38	0	1	159	8	1	4	277	180	1	3	395	2	1	3	513	26	1	3
42	158	0	1	160	38	1	3	278	107	1	2	396	52	0	3	514	66	0	2
43	65	1	2	161	10	0	3	279	18	1	3	397	21	1	1	515	64	1	2
44	126	0	1	162	8	0	3	280	85	1	3	398	3	1	3	516	57	1	2
45	10	1	1	163	6	1	3	281	61	1	3	399	16	1	3	517	84	0	2
46	2	1	3	164	28	0	1	282	129	0	3	400	9	0	3	518	4	1	3
47	6	1	3	165	10	1	2	283	13	1	3	401	11	1	3	519	82	0	2
48	5	0	3	166	99	0	3	284	37	1	3	402	8	1	3	520	2	1	2
49	14	1	1	167	38	1	3	285	112	0	3	403	30	1	1	521	18	1	2
50	15	1	2	168	40	1	2	286	22	1	3	404	2	0	3	522	82	0	2
51	114	0	2	169	47	0	2	287	65	0	3	405	4	1	4	523	9	1	2
52	13	1	2	170	11	1	2	288	9	1	3	406	61	0	1	524	7	1	2
53	10	1	2	171	60	0	3	289	37	1	3	407	8	1	3	525	80	0	3
54	14	0	2	172	47	0	3	290	62	1	1	408	18	0	3	526	35	1	2
55	62	1	1	173	11	1	2	291	39	1	2	409	23	1	2	527	51	1	2
56	34	1	3	174	17	1	4	292	30	1	2	410	2	1	3	528	25	1	1
57	37	0	1	175	35	1	2	293	42	1	2	411	25	1	1	529	30	1	2
58	245	0	1	176	30	1	4	294	18	1	3	412	20	0	1	530	4	1	2
59	65	1	1	177	19	1	3	295	25	1	3	413	23	1	2	531	1	1	2
60	15	0	2	178	201	0	1	296	70	1	3	414	43	1	2	532	7	1	2
61	2	1	2	179	23	1	1	297	86	1	2	415	1	1	1	533	27	0	3
62	70	1	1	180	41	1	3	298	87	0	2	416	86	0	1	534	26	0	2
63	38	1	2	181	35	1	1	299	33	1	3	417	38	1	2	535	2	1	2
64	8	1	2	182	40	0	1	300	11	1	3	418	112	0	2	536	114	0	1
65	180	0	3	183	15	1	2	301	24	1	3	419	15	1	2	537	17	0	2
66	74	0	1	184	56	1	1	302	36	1	1	420	105	0	1	538	73	0	3
67	10	0	2	185	11	1	2	303	1	1	2	421	115	1	2	539	31	0	1
68	46	1	2	186	50	1	2	304	35	1	2	422	46	0	1	540	63	1	2
69	44	1	1	187	125	0	4	305	52	0	1	423	47	0	1	541	40	0	2
70	10	1	2	188	29	1	2	306	31	1	2	424	2	1	2	542	13	1	2
71	25	1	1	189	19	1	3	307	22	1	3	425	5	0	2	543	10	1	2
72	129	0	1	190	31	1	3	308	1	1	3	426	4	1	2	544	8	1	2

i	X_i	δ	z_i	i	X_i	δ	z_i	i	X_i	δ	z_i	i	X_i	δ	z_i	i	X_i	δ	z_i
73	146	1	2	191	158	1	4	309	72	0	2	427	4	1	2	545	48	1	1
74	105	0	1	192	42	0	2	310	10	1	3	428	76	1	1	546	117	1	1
75	28	1	4	193	122	0	1	311	27	1	3	429	42	0	2	547	30	1	2
76	44	1	2	194	22	1	1	312	9	1	2	430	47	1	2	548	23	0	1
77	80	1	2	195	54	1	1	313	42	1	3	431	35	1	2	549	31	0	1
78	159	0	2	196	29	1	3	314	29	1	1	432	72	1	2	550	12	1	2
79	4	1	2	197	47	1	3	315	101	0	2	433	85	0	3	551	28	1	2
80	45	1	2	198	16	1	3	316	83	1	1	434	89	1	2	552	87	0	1
81	8	1	2	199	10	1	2	317	34	1	3	435	2	0	2	553	7	1	2
82	10	1	2	200	111	0	3	318	55	0	2	436	20	1	3	554	36	0	2
83	9	1	2	201	32	0	3	319	64	0	1	437	59	1	3	555	67	1	1
84	84	1	2	202	119	0	3	320	13	1	1	438	23	1	1	556	32	1	2
85	55	1	2	203	38	1	1	321	51	1	2	439	36	1	2	557	33	1	1
86	35	1	2	204	1	1	2	322	20	1	2	440	15	1	2	558	37	0	1
87	16	1	3	205	124	0	1	323	47	1	2	441	2	1	3	559	34	0	2
88	7	1	3	206	14	1	1	324	25	1	2	442	33	1	3	560	7	1	1
89	40	1	2	207	22	1	3	325	88	1	3	443	4	1	3	561	18	1	3
90	2	1	2	208	192	0	3	326	85	1	2	444	127	1	2	562	23	1	4
91	11	1	2	209	5	1	3	327	16	1	1	445	4	0	2	563	78	0	3
92	56	1	1	210	83	0	2	328	62	0	3	446	84	1	2	564	3	1	2
93	122	1	2	211	1	1	1	329	35	1	3	447	11	0	2	565	26	1	2
94	120	0	2	212	8	1	3	330	90	1	2	448	13	0	2	566	10	0	2
95	8	1	2	213	16	1	2	331	49	0	3	449	30	1	1	567	51	1	2
96	5	1	2	214	17	1	3	332	14	1	3	450	17	1	2	568	47	1	2
97	40	1	1	215	3	1	3	333	11	1	3	451	18	1	3	569	113	0	1
98	6	1	3	216	10	1	3	334	30	1	1	452	13	1	3	570	43	0	1
99	82	0	1	217	11	1	2	335	7	1	1	453	3	1	3	571	8	1	3
100	51	1	1	218	123	1	1	336	56	1	3	454	18	1	3	572	19	1	3
101	85	0	1	219	47	0	3	337	2	1	1	455	77	1	3	573	95	1	2
102	35	1	2	220	16	1	3	338	79	0	2	456	9	1	3	574	11	0	3
103	13	1	2	221	28	1	3	339	77	1	2	457	44	1	2	575	22	1	3
104	72	1	2	222	79	0	1	340	46	1	3	458	62	0	2	576	8	1	1
105	44	1	3	223	15	1	3	341	7	1	2	459	3	1	4	577	23	0	1
106	18	1	2	224	8	1	3	342	44	1	1	460	11	1	2	578	2	1	1
107	18	0	1	225	43	0	2	343	34	1	1	461	10	1	3	579	105	0	3

i	X_i	δ	z_i	i	X_i	δ	z_i	i	X_i	δ	z_i	i	X_i	δ	z_i	i	X_i	δ	z_i
108	94	0	1	226	8	1	3	344	57	1	3	462	10	1	2	580	2	1	3
109	25	1	1	227	12	1	2	345	37	1	1	463	60	1	2	581	6	1	3
110	88	0	3	228	173	0	2	346	5	1	4	464	6	0	2	582	10	1	3
111	11	1	3	229	18	1	3	347	92	0	3	465	32	1	1	583	23	1	3
112	79	0	2	230	23	0	2	348	46	0	3	466	5	1	3	584	22	1	3
113	147	0	1	231	3	1	3	349	20	1	3	467	20	1	2	585	45	1	2
114	17	1	2	232	42	1	2	350	11	1	2	468	23	1	1	586	6	1	2
115	64	0	3	233	23	1	1	351	96	0	2	469	29	1	3				
116	45	1	2	234	33	1	2	352	14	1	3	470	9	1	3				
117	28	1	3	235	37	1	1	353	13	1	2	471	6	1	3				
118	33	1	2	236	1	1	2	354	13	1	2	472	6	1	2				

Table A2.3. *The lifetimes of 586 lung cancer patients after operation, in months*

Bibliography

[AAL 78] AALEN O., "Nonparametric inference for the family of counting processes", *Annals of Statistics*, vol. 6, pp. 701–726, 1978.

[AGU 02] AGUIRRE-HERNANDEZ R., FAREWELL V., "A Pearson-type goodness-of-fit test for stationary and time-continuous Markov regression models", *Statistics in Medicine*, vol. 21, pp. 1899–1911, 2002.

[AKR 88] AKRITAS M., "Pearson-type goodness-of-fit tests: the univariate case", *Journal of the American Statistical Association*, vol. 83, pp. 222–230, 1988.

[AKR 97] AKRITAS M., TORBEYNS A., "Pearson-type goodness-of-fit tests for regression", *Canadian Journal of Statistics*, vol. 25, pp. 359–374, 1997.

[AND 93] ANDERSEN P., BORGAN O., GILL R. *et al.*, *Statistical Models Based on Counting Processes*, Springer-Verlag, New York, 1993.

[BAG 02] BAGDONAVIČIUS V., NIKULIN M., *Accelerated Life Models*, Chapman and Hall/CRC, Boca Raton, 2002.

[BAG 10a] BAGDONAVIČIUS V., KRUOPIS J., NIKULIN M., *Nonparametric Tests for Complete Data*, ISTE Ltd, London and John Wiley & Sons, New York, 2010.

[BAG 10b] BAGDONAVIČIUS V., KRUOPIS J., NIKULIN M., *Nonparametric Tests for Censored Data*, ISTE Ltd, London and John Wiley & Sons, New York, 2010.

[BAG 11] BAGDONAVIČIUS V., NIKULIN M., "Chi-squared goodness-of-fit test for right censored data", *International Journal of Applied Mathematics and Statistics*, vol. 24, pp. 30–50, 2011.

[BAG 13] BAGDONAVIČIUS V., LEVULIENE R., NIKULIN M., "Chi-squared goodness-of-fit tests for parametric accelerated failure time models", *Communications in Statistics – Theory and Methods*, vol. 42, 2013.

[BAR 73] BARR D., DAVIDSON T., "A Kolmogorov–Smirnov test for censored samples", *Technometrics*, vol. 15, no. 4, pp. 739–757, 1973.

[BEI 82] BEINICKE G., DZHAPARIDZE K., "On parameter estimation by the Davidon–Fletcher–Powell method", *Theory of Probability and Its Application*, vol. 27, pp. 374–380, 1982.

[BIL 68] BILLINGSLEY P., *Convergence of Probability Measures*, John Wiley & Sons, New York, 1968.

[BOL 75] BOLSHEV L., NIKULIN M., "One solution of the problem of homogeneity", *Serdika, B'lgarsko Mathematichesko Spicanie*, vol. 1, pp. 104–109, 1975.

[BOL 78] BOLSHEV L., MIRVALIEV M., "Chi-square goodness-of-fit test for the Poisson, binomial, and negative binomial distributions", *Theory of Probability and Its Applications*, vol. 23, no. 2, pp. 461–474, 1978.

[BOL 83] BOLSHEV L., SMIRNOV N., *Tables of Mathematical Statistics*, Nauka, Moscow, 1983.

[BOR 84] BORGAN O., "Maximum likelihood estimation in parametric counting process models, with applications to censored failure time data", *Scandinavian Journal of Statistics*, vol. 11, pp. 1–16, 1984.

[CAS 11] CASTRO-KURISS C., "On a goodness-of-fit test for censored data from a location-scale distribution with applications", *Chilean Journal of Statistics*, vol. 2, no. 1, pp. 115–136, 2011.

[CHE 54] CHERNOFF H., LEHMANN E., "The use of maximum likelihood estimates in χ^2 tests for goodness of fit", *Annals of Mathematical Statistics*, vol. 25, pp. 579–586, 1954.

[CHI 71] CHIBISOV D., "Certain chi-square type tests for continuous distributions", *Theory of Probability and Its Applications*, vol. 16, no. 1, pp. 3–21, 1971.

[CHI 11a] CHIMITOVA E., NIKULIN M., LEMESHKO B. *et al.*, "Nonparametric goodness-of-fit tests for censored data", *Proceedings of the 7th International Conference on Mathematical Methods in Reliability: Theory, Methods and Applications*, Beijing, China, pp. 817–823, 2011.

[CHI 11b] CHIMITOVA E., LIERO H., VEDERNIKOVA M., "Application of classical Kolmogorov, Cramer-von Mises-Smirnov and Anderson–Darling tests for censored samples", *Proceedings of the International Workshop on Applied Methods of Statistical Analysis*, Novosibirsk, Russia, pp. 176–185, 2011.

[CHI 11c] CHIMITOVA E., TSIVINSKAYA A., "Simulation study for the NRR chi-square test of goodness-of-fit for censored data", *Proceedings of the International Workshop on Applied Methods of Statistical Analysis*, Novosibirsk, Russia, pp. 44–52, 2011.

[CHI 14] CHIMITOVA E., LEMESHKO B., "A comparative analysis of some chi-squared goodness-of-fit tests for censored data", in COUALLIER V. *et al.* (eds), *Statistical Models and Methods for Reliability and Survival Analysis*, ISTE Ltd, London and John Wiley & Sons, New York, 2014.

[COH 91] COHEN A., *Truncated and Censored Samples: Theory and Applications*, Marcel Dekker, New York, 1991.

[CRA 46] CRAMER H., *Mathematical Methods of Statistics*, Princeton University Press, Princeton, 1946.

[D'AG 86] D'AGOSTINO R., STEPHENS M., *Goodness of Fit Techniques*, Marcel Dekker, New York, 1986.

[DEN 79] DENISOV V., LEMESHKO B., "Optimal grouping in the analysis of experimental data", *Measuring Information Systems*, Novosibirsk, pp. 5–14, 1979.

[DEN 89] DENISOV V., LEMESHKO B., "Optimal grouping in estimation and tests of goodness-of-fit hypotheses", *Wissenschaftliche Schriftenreihe der Technishen universitat Karl-Marx-Stadt.*, vol. 10, pp. 63–81, 1989.

[DEN 04] DENISOV V., EGER K.-H., LEMESHKO B. *et al.*, *Design of Experiments and Statistical Analysis for Grouped Observations*, NSTU Publishing House, Novosibirisk, 2004.

[DRO 88] DROST F., *Asymptotics for Generalized Chi-Squared Goodness-of-Fit Tests*, vol. 48, Center for Mathematics and Computer Sciences CWI Tract, Amsterdam, 1988.

[DRO 89] DROST F., "Generalized chi-square goodness-of-fit tests for location-scale models when the number of classes tends to infinity", *Annals of Statistics*, vol. 17, no. 3, pp. 1285–1300, 1989.

[DUD 79] DUDLEY R., "On χ^2 tests of composite hypothesis", *Probability Theory*, vol. 5, pp. 75–87, 1979.

[DZH 74] DZHAPARIDZE K., NIKULIN M., "On a modification of the standard statistic of Pearson", *Theory of Probability and Its Applications*, vol. 19, no. 4, pp. 851–852, 1974.

[DZH 82] DZHAPARIDZE K., NIKULIN M., "Probability distributions of the Kolmogorov and omega-square statistics for continuous distributions with shift and scale parameters", *Journal of Soviet Mathematics*, vol. 20, no. 3, pp. 2147–2163, 1982.

[DZH 83] DZHAPARIDZE K., "On iterative procedures of asymptotic inference", *Statistics Neerlandica*, vol. 37, pp. 181–189, 1983.

[GER 79] GERLACH B., "A consistent correlation-type goodness-of-fit test with application to the two-parameter weibull distribution", *Series Statistics*, vol. 10, no. 3, pp. 427–452, 1979.

[GRA 85] GRAY R., PIERCE D., "Goodness of fit tests for censored survival data", *Annals of Statistics*, vol. 13, pp. 552–563, 1985.

[GRE 88] GREENWOOD P., NIKULIN M., "Application of tests of chi-square type", *Journal of Soviet Mathematics*, vol. 43, no. 6, pp. 2776–2791, 1988.

[GRE 96] GREENWOOD P., NIKULIN M., *A Guide to Chi-Squared Testing*, John Wiley & Sons, New York, 1996.

[HAB 86] HABIB M., THOMAS D., "Chi-square goodness-of-fit tests for randomly censored data", *Annals of Statistics*, vol. 14, pp. 759–765, 1986.

[HJO 90] HJORT N., "Goodness of fit tests in models for life history data based on cumulative hazard rates", *Annals of Statistics*, vol. 18, no. 3, pp. 1221–1258, 1990.

[HOL 92] HOLLANDER M., PEÑA E., "A chi-squared goodness-of-fit test for randomly censored data", *Journal of the American Statistical Association*, vol. 87, pp. 458–463, 1992.

[IBR 81] IBRAGIMOV I., HAS'MINSKII R., *Statistical Estimation, Asymptotic Theory*, Springer, New York, 1981.

[KAL 85] KALBFLEISCH J., LAWLESS J., "The analysis of panel data under a Markov assumption", *Journal of the American Statistical Association*, vol. 80, pp. 863–871, 1985.

[KAP 58] KAPLAN E., MEIER P., "Nonparametric estimation from incomplete observations", *Journal of the American Statistical Association*, vol. 53, pp. 457–481, 1958.

[KEC 02] KECECIOGLU D., *Reliability Engineering Handbook*, vol. 1, DEStech Publications, Lancaster, 2002.

[KIM 93] KIM J., "Chi-square goodness-of-fit tests for randomly censored data", *Annals of Statistics*, vol. 21, no. 3, pp. 1621–1639, 1993.

[KLE 03] KLEIN J., MOESCHBERGER M., *Survival Analysis: Techniques for Censored and Truncated Data*, 2nd ed., Springer, New York, 2003.

[KOU 84] KOUL H., "Tests of goodness-of-fit in linear regression", *Colloquia Mathematica Societatis János Bolyai*, vol. 45, pp. 279–315, 1984.

[KOZ 76] KOZIOL J., GREEN S., "A Cramer-von Mises statistic for randomly censored data", *Biometrika*, vol. 63, pp. 465–474, 1976.

[LAW 03] LAWLESS J., *Statistical Models and Methods for Lifetime Data*, Wiley, Chichester, 2003.

[LEC 56] LECAM L., "On the asymptotic theory of estimation and testing hypotheses", *Proceedings of the Third Berkeley Symposium on Mathematical Statistics and Probability*, Berkeley, CA, pp. 129–156, 1956.

[LEC 60] LECAM L., "Locally asymptotically normal families of distributions", *University of California Publication in Statistics*, vol. 3, no. 2, pp. 37–98, 1960.

[LEC 83] LECAM L., MAHAN C., SINGH A., "An extension of a theorem of H. Chernoff and E.L. Lehmann", in RIZVI H. *et al.* (eds), *Recent Advances in Statistics*, Academic Press, 1983.

[LEM 98] LEMESHKO B., "Asymptotically optimal grouping of observations in goodness-of-fit tests", *Industrial Laboratory*, vol. 64, no. 1, pp. 59–67, 1998.

[LEM 00] LEMESHKO B., CHIMITOVA E., "Maximization of the power of χ^2 tests", *Papers of Siberian Branch of Academy of Science of Higher Education*, vol. 2, pp. 53–61, 2000.

[LEM 01] LEMESHKO B., POSTOVALOV S., CHIMITOVA E., "On statistic distributions and the power of the Nikulin χ^2 test", *Industrial Laboratory*, vol. 67, no. 3, pp. 52–58, 2001.

[LEM 05] LEMESHKO B., LEMESHKO S., "Comparative analysis of tests for verifying deviation of a distribution from normal", *Metrologiya*, no. 2, pp. 3–24, 2005.

[LEM 09a] LEMESHKO B., LEMESHKO S., "Distribution models for nonparametric tests for fit in verifying complicated hypotheses and maximum-likelihood estimators. Part 1", *Measurement Techniques*, vol. 52, pp. 555–565, 2009.

[LEM 09b] LEMESHKO B., LEMESHKO S., "Models for statistical distributions in nonparametric fitting tests on composite hypotheses based on maximum-likelihood estimators. Part II", *Measurement Techniques*, vol. 52, pp. 799–812, 2009.

[LEM 09c] LEMESHKO B., LEMESHKO S., POSTOVALOV S., "Comparative analysis of the power of goodness-of-fit tests for near competing hypotheses. Part I. Verification of simple hypotheses", *Journal of Applied and Industrial Mathematics*, vol. 3, no. 4, pp. 462–475, 2009.

[LEM 10a] LEMESHKO B., CHIMITOVA E., PLESHKOVA T., "Testing simple and composite goodness-of-fit hypotheses by censored samples", *Nauchniy Vestnik NGTU*, vol. 41, no. 4, pp. 13–28, 2010.

[LEM 10b] LEMESHKO B., LEMESHKO S., POSTOVALOV S., "Comparative analysis of the power of goodness-of-fit tests for near competing hypotheses. Part II. Verification of complex hypotheses", *Journal of Applied and Industrial Mathematics*, vol. 4, no. 1, pp. 79–93, 2010.

[LEM 11a] LEMESHKO B., LEMESHKO S., AKUSHKINA K. *et al.*, "Inverse Gaussian model and Its applications in reliability and survival analysis", in RYKOV V., BALAKRISHNAN N., NIKULIN M. (eds.), *Mathematical and Statistical Models and Methods in Reliability. Applications to Medicine, Finance, and Quality Control*, Birkhäuser, Boston, 2011.

[LEM 11b] LEMESHKO B., LEMESHKO S., POSTOVALOV S. *et al.*, *Statistical Data Analysis, Simulation and Research of Probabilistic Regularities. Computer Approach*, NSTU Publishing House, Novosibirsk, 2011.

[LEM 15a] LEMESHKO B., "Chi-square-type tests for verification of normality", *Measurement Techniques*, vol. 58, no. 6, pp. 581–591, 2015.

[LEM 15b] LEMESHKO B., *Criteria for Testing the Deviation of Distribution from the Normal Law. Application Guide*, INFRA-M, Moscow, 2015.

[LI 93] LI G., DOSS H., "Generalized Pearson–Fisher chi-square goodness-of-fit tests, with applications to models with life history data", *Annals of Statistics*, vol. 21, pp. 772–797, 1993.

[LOY 80] LOYNES R., "The empirical distribution function of residuals from generalized regression", *Annals of Statistics*, vol. 8, pp. 285–298, 1980.

[MAN 73] MANN N., SCHEUER E., FERTIG K., "A new goodness-of-fit test for the two parameter weibull or extreme – value distribution with unknown parameters", *Communications in Statistics*, vol. 2, pp. 283–400, 1973.

[MIR 89] MIRVALIEV M., NIKULIN M., "About homogeneity of two samples", *Proceedings of the International Vilnius Conference on Probability Theory and Mathematical Statistics*, pp. 40–41, Vilnius, 1989.

[MUK 77] MUKANTSEVA L., "Testing normality in one-dimensional and multi-dimensional linear regression", *Theory of Probability and Its Applications*, vol. 22, pp. 591–602, 1977.

[NAI 81] NAIR V., "Plots and tests for goodness of fit with randomly censored data", *Biometrika*, vol. 68, pp. 99–103, 1981.

[NEL 90] NELSON W., *Accelerated Testing: Statistical Models, Test Plans, and Data Analysis*, Wiley, New York, 1990.

[NIK 73a] NIKULIN M., "Chi-square test for continuous distributions", *Theory of Probability and Its Applications*, vol. 18, pp. 638–639, 1973.

[NIK 73b] NIKULIN M., "Chi-square test for continuous distributions with shift and scale parameters", *Theory of Probability and Its Applications*, vol. 18, pp. 559–568, 1973.

[NIK 73c] NIKULIN M., "Chi-square test for normality", *Proceedings of the International Vilnius Conference on Probability Theory and Mathematical Statistics*, Vilnius, pp. 119–122, 1973.

[NIK 74] NIKULIN M., "On a quantile test", *Theory of Probability and Its Applications*, vol. 19, no. 2, pp. 410–413, 1974.

[NIK 79a] NIKULIN M., "Applications of the model of two-factor analysis of variance without interaction", *Journal of Soviet Mathematics*, vol. 25, no. 3, pp. 1196–1207, 1979.

[NIK 79b] NIKULIN M., "Hypothesis testing for a parameter difference in binomial distributions", *Theory of Probability and Its Applications*, vol. 24, no. 2, pp. 392–396, 1979.

[NIK 83] NIKULIN M., YUSAS J., "Distributions of a χ^2 statistic with additional observations used to estimate the unknown parameter", *Journal of Soviet Mathematics*, vol. 23, no. 3, pp. 2290–2295, 1983.

[NIK 89] NIKULIN M., VOINOV V., "A chi-square goodness-of-fit test for exponential distributions of the first order", in KALASHNIKOV V., ZOLOTAREV V. (eds), *Stability Problems for Stochastic Models*, Springer, 1989.

[NIK 90a] NIKULIN M., "L'exaustivité, l'estimation sans biais et le test d'ajustement du chi-deux", *Statistique des processus en milieu médical*, vol. 5, pp. 176–198, 1990.

[NIK 90b] NIKULIN M., VOINOV V., "A chi-square goodness-of-fit test for multivariate exponential distributions", in KUBIK S., VISEK J.A. (eds.), *Information Theory, Statistical Decision Functions and Random Processes*, Springer, Berlin, 1990.

[NIK 91] NIKULIN M., "Some recent results on chi-squared tests", *Queen's Papers in Pure and Applied Mathematics*, vol. 86, 1991.

[NIK 99] NIKULIN M., SOLEV V.N., "Chi-squared goodness of fit test for Doubly censored data with applications in survival analysis and reliability", in IONESCU D.C., LIMNIOS N. (eds.), *Statistical and Probabilistic Models in Reliability*, Birkhauser, Boston, 1999.

[PEN 98a] PENA E., "Smooth goodness-of-fit tests for composite hypothesis in hazard based models", *Annals of Statistics*, vol. 26, pp. 1935–1971, 1998.

[PEN 98b] PENA E., "Smooth goodness-of-fit tests for composite hypothesis in hazard based models", *Journal of the American Statistical Association*, vol. 93, no. 442, pp. 673–692, 1998.

[PIE 79] PIERCE D., KOPECKY K., "Testing goodness of fit for the distribution of errors in regression models", *Biometrika*, vol. 66, pp. 1–6, 1979.

[RAO 74] RAO K., ROBSON D., "A chi-squared statistic for goodness-of-fit tests within the exponential family", *Communications in Statistics*, vol. 3, pp. 1139–1153, 1974.

[REI 04] REINEKE D., CROWN J., "Estimation of hazard, density and survival functions for randomly censored data", *Journal of Applied Statistics*, vol. 31, pp. 1211–1225, 2004.

[ROY 11] ROYSTON P., LAMBERT P., *Flexible Parametric Survival Analysis Using Strata: Beyond the Cox Model*, A Strata Press Publications, Texas, 2011.

[SAS 08] SASCO A., NIKULIN M., "Flexible regression models for dynamic prediction of survival in elderly in presence of cancer or other diseases", *Advances in Gerontology*, vol. 21, pp. 41–48, 2008.

[TIT 08] TITMAN A., SHARPLES L., "A general goodness-of-fit test for Markov and hidden Markov models", *Statistics in Medicine*, vol. 27, pp. 2177–2195, 2008.

[VAN 98] VAN DER VAART A., *Asymptotic Statistics*, Cambridge University Press, 1998.

[VOI 93a] VOINOV V., NIKULIN M., *Unbiased Estimators and Their Applications. 1: Univariate Case*, Kluwer Academic Publishers, Dordrecht, 1993.

[VOI 93b] VOINOV V., NIKULIN M., "A chi-squared goodness-of-fit test for one- and multidimensional distributions", *Journal of Soviet Mathematics*, vol. 9, pp. 62–79, 1993.

[VOI 96] VOINOV V., NIKULIN M., *Unbiased Estimators and Their Applications. 2: Multivariate Case*, Kluwer Academic Publishers, Dordrecht, 1996.

[VOI 06] VOINOV V., "On optimality of the Rao–Robson–Nikulin test (in Russian)", *Industrial Laboratory*, vol. 72, pp. 65–70, 2006.

[VOI 07] VOINOV V., NIKULIN M., PYA N., "Independent χ_1^2 distributed in the limit components of some chi-squared tests", in SKIADIS C. (ed.), *Recent Advance in Stochastic Modelling and Data Analysis*, World Scientific, London, 2007.

[VOI 09] VOINOV V., PYA N., ALLOYAROVA R., "A comparative study of some modified chi-squared tests", *Communications in Statistics-Simulation and Computation*, vol. 38, pp. 355–367, 2009.

[VOI 13] VOINOV V., NIKULIN M., BALAKRISHNAN N., *Chi-Squared Goodness of Fit Tests with Applications*, Academic Press, Cambridge, 2013.

[ZAC 71] ZACKS S., *The Theory of Statistical Inference*, Wiley, New York, 1971.

[ZHA 99] ZHANG B., "A chi-squared goodness of-fit test for logistic regression models based on case-control data", *Biometrika*, vol. 86, pp. 531–539, 1999.

[ZHA 02] ZHANG J., "Powerful goodness-of-fit tests based on the likelihood ratio", *Journal of the Royal Statistical Society: Series B*, vol. 64, no. 2, pp. 281–294, 2002.

[ZHA 05] ZHANG J., "Likelihood-ratio tests for normality", *Computational Statistics & Data Analysis*, vol. 49, no. 3, pp. 709–721, 2005.

Index

P, R, S, T, W, Z

Other titles from

iSTE

in

Mathematics and Statistics

2017

CELANT Giorgio, BRONIATOWSKI Michel
Interpolation and Extrapolation Optimal Designs 2: Finite Dimensional General Models

D'AMICO Guglielmo, DI BIASE Giuseppe, JANSSEN Jacques, MANCA Raimondo
Semi-Markov Migration Models for Credit Risk (Stochastic Models for Insurance Set – Volume 1)

HARLAMOV Boris
Stochastic Analysis of Risk and Management (Stochastic Models in Survival Analysis and Reliability Set – Volume 2)

SIMON Jacques
Banach, Fréchet, Hilbert and Neumann Spaces (Analysis for PDEs Set – Volume 1)

2016

CELANT Giorgio, BRONIATOWSKI Michel
Interpolation and Extrapolation Optimal Designs 1: Polynomial Regression and Approximation Theory

CHIASSERINI Carla Fabiana, GRIBAUDO Marco, MANINI Daniele
Analytical Modeling of Wireless Communication Systems (Stochastic Models in Computer Science and Telecommunication Networks Set – Volume 1)

GOUDON Thierry
Mathematics for Modeling and Scientific Computing

KAHLE Waltraud, MERCIER Sophie, PAROISSIN Christian
Degradation Processes in Reliability
(Mathematial Models and Methods in Reliability Set – Volume 3)

KERN Michel
Numerical Methods for Inverse Problems

RYKOV Vladimir
Reliability of Engineering Systems and Technological Risks
(Stochastic Models in Survival Analysis and Reliability Set – Volume 1)

2015

DE SAPORTA Benoîte, DUFOUR François, ZHANG Huilong
Numerical Methods for Simulation and Optimization of Piecewise Deterministic Markov Processes

DEVOLDER Pierre, JANSSEN Jacques, MANCA Raimondo
Basic Stochastic Processes

LE GAT Yves
Recurrent Event Modeling Based on the Yule Process
(Mathematical Models and Methods in Reliability Set – Volume 2)

2014

COOKE Roger M., NIEBOER Daan, MISIEWICZ Jolanta
Fat-tailed Distributions: Data, Diagnostics and Dependence (Mathematical Models and Methods in Reliability Set – Volume 1)

MACKEVIČIUS Vigirdas
Integral and Measure: From Rather Simple to Rather Complex

PASCHOS Vangelis Th
Combinatorial Optimization – 3-volume series – 2nd edition
Concepts of Combinatorial Optimization / Concepts and Fundamentals – volume 1
Paradigms of Combinatorial Optimization – volume 2
Applications of Combinatorial Optimization – volume 3

2013

COUALLIER Vincent, GERVILLE-RÉACHE Léo, HUBER Catherine, LIMNIOS Nikolaos, MESBAH Mounir
Statistical Models and Methods for Reliability and Survival Analysis

JANSSEN Jacques, MANCA Oronzio, MANCA Raimondo
Applied Diffusion Processes from Engineering to Finance

SERICOLA Bruno
Markov Chains: Theory, Algorithms and Applications

2012

BOSQ Denis
Mathematical Statistics and Stochastic Processes

CHRISTENSEN Karl Bang, KREINER Svend, MESBAH Mounir
Rasch Models in Health

DEVOLDER Pierre, JANSSEN Jacques, MANCA Raimondo
Stochastic Methods for Pension Funds

2011

MACKEVIČIUS Vigirdas
Introduction to Stochastic Analysis: Integrals and Differential Equations

MAHJOUB Ridha
Recent Progress in Combinatorial Optimization – ISCO2010

RAYNAUD Hervé, ARROW Kenneth
Managerial Logic

2010

BAGDONAVIČIUS Vilijandas, KRUOPIS Julius, NIKULIN Mikhail
Nonparametric Tests for Censored Data

BAGDONAVIČIUS Vilijandas, KRUOPIS Julius, NIKULIN Mikhail
Nonparametric Tests for Complete Data

IOSIFESCU Marius *et al.*
Introduction to Stochastic Models

VASSILIOU PCG
Discrete-time Asset Pricing Models in Applied Stochastic Finance

2008

ANISIMOV Vladimir
Switching Processes in Queuing Models

FICHE Georges, HÉBUTERNE Gérard
Mathematics for Engineers

HUBER Catherine, LIMNIOS Nikolaos *et al.*
Mathematical Methods in Survival Analysis, Reliability and Quality of Life

JANSSEN Jacques, MANCA Raimondo, VOLPE Ernesto
Mathematical Finance

2007

HARLAMOV Boris
Continuous Semi-Markov Processes

2006

CLERC Maurice
Particle Swarm Optimization

Printed and bound by CPI Group (UK) Ltd, Croydon, CR0 4YY

15/09/2026

14962355-0001